TECHNOLOGY, INNOVATION and POLICY 13

Series of the Fraunhofer Institute
for Systems and Innovation Research ISI

TECHNOLOGY, INNOVATION and POLICY

Series of the Fraunhofer Institute
for Systems and Innovation Research ISI

Kerstin Cuhls · Knut Blind
Hariolf Grupp

Innovations for our Future

Delphi '98:
New Foresight on Science and Technology

With Contributions of
Harald Bradke · Carsten Dreher · Dirk M. Harmsen
Harald Hiessl · Bärbel Hüsing · Gerhard Jaeckel
Ulrich Schmoch · Peter Zoche

With 66 Figures and 29 Tables

Springer-Verlag Berlin
Heidelberg GmbH

Dr. Kerstin Cuhls
Dr. Knut Blind
Professor Dr. Hariolf Grupp
Fraunhofer Institute for Systems
and Innovation Research ISI
Breslauer Straße 48
76139 Karlsruhe
Germany
cu@isi.fhg.de
kb@isi.fhg.de
hariolf.grupp@isi.fhg.de

Secretariat:
Renate Klein
Chris Mahler-Johnstone

ISSN 1431-9667

ISBN 978-3-7908-1434-7   ISBN 978-3-642-57472-6 (eBook)
DOI 10.1007/978-3-642-57472-6

Cataloging-in-Publication Data applied for
Die Deutsche Bibliothek – CIP-Einheitsaufnahme
Cuhls, Kerstin: Innovations for our future: Delphi 98: new foresight on science and technology; with
29 tables / Kerstin Cuhls; Knut Blind; Hariolf Grupp. With contributions of Harald Bradke ...
Heidelberg; New York: Physica-Verl., 2002
   (Technology, innovation and policy; 13)

© Springer-Verlag Berlin Heidelberg 2002
Originally published by Physica-Verlag Heidelberg 2002

Softcover design: Erich Kirchner, Heidelberg
SPIN 10850774   88/2202-5 4 3 2 1 0 – Printed on acid-free paper

Acknowledgements

The Delphi '98 study about the global development of science and technology was a highly risky project, as it experimented with the Delphi methodology and was carried out on such a large scale for the first time. The survey was conducted by the Fraunhofer Institute for Systems and Innovation Research on behalf of the Federal German Ministry for Education and Research. The authors wish to thank the Ministry for their confidence in this activity and their sponsorship of it.

But the study would not have been possible without the commitment of the steering committee, the experts groups who prepared the questionnaire topics, and the experts who filled in such a long questionnaire concerning future innovations. We thank them very much for not giving up, although some of them wondered why the same questionnaire had to be filled in twice. No, we do not think the experts are stupid, on the contrary, the crucial point of the Delphi methodology is that we do not know the future and therefore gather the information twice. Maybe they also learned something: about future innovations, about the opinions of other experts and if there is consensus concerning these developments – or not. As it may be possible to update the Delphi '98 in two or three years' time, we hope that we will find similarly enthusiastic experts again. We hope that we can count on their cooperation once more.

We also thank our international cooperation partners, especially at the National Institute for Science and Technology Policy, for the exchange of experiences and data which made an international comparison possible.

ISI had to suffer from the study, too. We thank all colleagues here for their cooperation and understanding. The persons responsible for the different fields must be mentioned: Harald Bradke, Carsten Dreher, Dirk M. Harmsen, Harald Hiessl, Bärbel Hüsing, Gerhard Jaeckel, Ulrich Schmoch, and Peter Zoche. They checked the structures of the Delphi fields, the single topics, made analyses, and also contributed to this volume. Renate Klein, Chris Mahler-Johnstone, Rebecca Rangnow, and Edeltraud Geibel prepared workshops, wrote texts, organised address databases, sent out thousands of letters and questionnaires, corrected texts, formated tables and figures. We know that without such support, not forgetting that of our students, a study like this is impossible. And the whole ISI had to suffer in February 1998 after the press release: for one week, the main telephone lines were all engaged because so many people had questions about the study or wanted to order it. Our administration was unprepared to sell so many reports (this was a sudden decision by the BMBF) – an unusual job for a research institution, and we did not expect to be inundated with orders. No one complained about this, most colleagues still smiled!

Last but not least, we have to thank our families and friends who had to wait for us so often: first, because we were conducting the study, then, because we were writing so many reports and articles – and for the last two years travelling around the world to present the data and exchange experiences on an international level.

June 2001

K. Cuhls
K. Blind
H. Grupp

Contents

List of Tables

List of Figures

1 History and Basic Methods

1.1 What can we know about the future?

How will the world develop? What lies ahead? What are the methods to find it out? What role will innovations play? Will we like them? Or will we need to intervene? These are questions affecting the way we shape our future, and answers to them can be sought in a number of ways: from horoscopes, from calculations, based on "feelings", or with the aid of modern, social science methods. The latter includes the systematic mobilization of the experiential knowledge of appropriate experts. If – as is the case with this book – the topic concerns the future of science and technology, it is the researchers themselves whose opinions are desired. Perhaps personal predicting lies behind this or serves primarily one's own interests rather than general predictions of the future. This is something we shall see in the future. But can the researchers who work on projects all be wrong?

This book about developments in the area of science and technology represents a combination of the general and the specific, of what we know about future developments and about the unknown. The analyses cover a large spectrum from the trivial (for the specialist) to the purely Utopian (for the "perennial dweller in the past"). After all, as Karl Popper notes: "We can know nothing of the future, otherwise we would know it...".

Delphi '98 was the result of previous experiences. This book seeks to explain how and why the study was carried out. The first part of the book (methodology) is designed for those who wish to learn more about the methods and results: How are such assessments arrived at? And what details make them up? This is followed by a look at megatrends and the more significant future tendencies. We will also attempt to give some order to the "jumbled perspective": results will be summarised by specific fields. Examples describing how the results can be utilised round out the second part of the study.

We first want to start with hindsight and consider briefly what the function of the priestess in the old Delphi temple was, whether historical research found any impacts on politics or society in those days and whether there was a lasting impact on the progress of mankind in prehistory. We will then pose the same queries to our present society, *"what is"* or better *"what could be* the function of technology foresight on our economy, policy and technology development"?

The chapters 3, 4 5, 6 and 7 give empirical results from the study Delphi '98. The last chapters mention the major methodological and empirical problems of a study like this, and give some results as an outlook on the future of foresight itself. But as good presentations start with a look back into the past, we will follow this tradition.

1.2   On the history of modern foresight

Thinking about the future and future events has a long history. People at all times wanted to know what lies ahead. That was the basis for the "success" of the Greek oracles in ancient times. That is also one of the reasons why horoscopes and fore-tellers always find their business. But the image on the future as well as the way of thinking in future dimensions changed over time. There were times when religious beliefs dictated or underlined the understanding of the future. In Europe, it was Christian thinking that influenced the "philosophy" - in other cultures different ways of thinking about the future developed. There were times, when even "scientific" results were ignored, because they offended the existing way of thinking in general and the thinking about future developments.

In these contexts, there was always the view to "the future" and the imagination of things to come. In general, the historical point of view was that there is the presence (now) and there is something coming, which is *the* future and only *one* future seems to be possible. Some called it fate, and most religions underlined this belief in the single possible future. It was mainly operations research that – for the first time – explicitly formulated the fact that the future can be shaped by the actions of today. These approaches already declared a kind of social engineering possible (Helmer 1966). Coming from the technical engineering angle, an attempt was made to translate the findings to social sciences and to "construct" whole societies by ap-plying scientific methods.[1]

The main initial work was performed at the RAND corporation, Santa Monica, California, in the years following 1948, the pioneers being A. Kaplan (1950), O. Helmer (1959; 1964; 1984), N. Rescher (1959), N. Dalkey (1969), Th. Gordon (1964) and others. "Forecasting", as it was known then, was motivated by Vanevar Bush's book "Science the endless frontier" (Bush 1945), advocating the transfor-mation of the US military economy research during World War II (e.g. the Man-hattan project) into long-term civilian research and commercial exploitation. The early attempts were also spurred by the amazing scientific successes of the Soviet "planned economy" (e.g., the hydrogen bomb or the launch of the Sputnik). In the

---

[1]   This does not mean that there were no visions or models of future societies available, but there was no real attempt to make them scientifically operable until that time.

context of forecasting work at RAND, a new innovation economics also developed (including work by Arrow, Winter, Nelson et al.; compare Hounshell, 1996).

Methodological starting points were systems analysis, operations research and comparable procedures. After early successes, many serious misconceptions of what "forecasting" ought to be arose. In the sixties and early seventies, the mechanical "prognosis" or "trend prediction" type of work based on "linear", i. e. sequential, models ceased to look interesting and the related forecasting activities fell into oblivion. This coincided with the end of the long growth euphoria following the War heralded by the first oil price adjustment; or the "Limits to Growth" report of the Club of Rome (Meadows et al. 1972). Although the "linear" models of thought were discarded (e.g. by the project "Hindsight", Sherwin/ Isenson 1967), some science policy communities further supported them for their legitimating power on research spending with no priorities (e.g. the project TRACES by the NSF 1968).

With the new evolutionary economics coming up based on selection procedures and the notion of variety generation by new products, and the sociology of science working on the functions of social systems in science as opposed to technology or the economy emphasising the "bounds of rationality" and "negotiating systems", it became clear that there may be a new, different use of forecasting methods. Martin and Irvine (1984) coined the term "foresight" and pointed to the communication or procedural power of it. The modern perception is that the actions of social systems, in particular science communities, cannot be predicted in terms of "natural" laws, and that future events in science and technology cannot be determined by extrapolation, but are shaped by these communities and a negotiating system. Martin and Johnston therefore now write on "Wiring up the national innovation system" (Martin/ Johnston 1999).

However, this present understanding of foresight was available in the literature from the very beginning and, though less well pronounced than nowadays, may have been found already in one of the earliest papers in the field:

"Policy making rests in part on anticipation of the future (...) and of the consequences of and responses to alternative lines of action. Many policy decisions require *foreknowledge* of events which cannot be forecast either by strict causal chains (...) or by stable statistical regularities (...)." [Kaplan et al. 1950, p.93].

Even the forerunner of the term "foresight", "foreknowledge", was coined in 1950. "Verification" or "falsification" of foresight results are, thus, meaningless ends for the early forecasters. But the full recognition of what this means for the Delphi technique as a tool in policy making was unfolded only later, in the 1980s, and based on the Japanese usage of it.

There is constant temptation for foresight to restrict itself to describing the potential supply of scientific and technical solutions and the study of their impact. However, it must do far more than depict the supply factors. The potentials and the risks of technology in the future depend just as much on the pressure of the social, ecological and economic problems expected to arise and make important demands on science and technology. For this reason, any discussion of problems must focus increasingly on factors relating to demand. How one might determine which basic values for innovation activities might be adopted worldwide in the medium and long-term perspectives, and forecast the resulting problems, of course, has no satisfactory empirical answer.

Because of many supply-demand mismatches, initially euphoric expectations of a new technology (mostly on the part of the scientific community) tend to be followed by increasingly cautious developmental phases before the market is finally satisfied. The use or rejection of innovative products often leads to new demands on research and technology, which is why it generally makes sense to speak of "feedback processes". Foresight has to incorporate aspects of industrial research and pure research, and consideration must also be given to institutional support. These deliberations also call into question the possible expectation that a technology needs no more than a single action to regulate its impacts. Any hopes of being able to drop the accompanying pure research once the applied objectives have been achieved, will meet with disappointment; tomorrow's science-based technology is shaped continuously through targeted basic research.

1.3   From forecasting to foresight

Single forecasting methods were developed that were supposed to be able to "explore" the future. Even the first attempts of scenarios by Herman Kahn (1967) were written with these connotations and therefore not different options of the future but only one single possibility for each option was described. As there is only a single present, only one future can occur. In fact, that is right to a certain extent: although Einstein explained the relativity of time, there will only be *one* present (e.g. experienced by me as a person), which was previously the future (fig. 1.3-1). So much for the underlying philosophical thinking.

But we have to admit that (at least at the moment) we cannot *explore* the future. The future is unknown to us – more or less. There will always be things that we can relatively exactly anticipate – and others that are surprising. Therefore, from the point of view of the present, there are always alternatives, which means different futures or future options from which only one option is realised as "the present" and later on referred to as "the past".

Figure 1.3-1:   Traditional view – single future

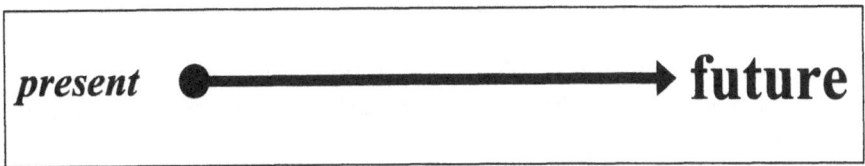

present ●————————————➔ **future**

With the scenario approaches developed at the end of the 60s, this image of "the future" was already taken into account. But it was expected to be able to filter out that single option which becomes reality, or to stress it: prediction was thought to be possible. Other methodologies also made use of this knowledge or considered their "single future approaches" as taking out one of the most promising, probable or wishful options (like in the Delphi methods). Thus, it was a backlash when the 1973 oil shock was not "predicted"[2]. As "right" prediction was the only criteria for forecasting and futures research in general (Sackman 1975 also made this mistake), forecasting fell into oblivion and was neglected by many decision-makers, planners and politicians.

At least in Europe, it took another 20 years to get its reputation back – with the changing label of "foresight". But as already mentioned in chapter 1.2, there is a difference in the understanding of forecasting in contrast to foresight. In this chapter, a very simple approach is made to explain *some* of the differences between forecasting, foresight and planning.

Forecasting is the look into the short-, medium- or long-term future with the means of scientific methodology, which can vary according to the areas of research or the questions posed concerning the future. Forecasting does not necessarily mean the belief in the prediction or the belief in the predictability of the future in general. It can be the view on one single future (see fig. 1.3-1), but it can also be the view on different futures (fig. 1.3-2). Figure 1.3-2 shows that many options are possible, and forecasting methodologies are used to try and identify them. But the success is unknown in advance. Sometimes, there are many more and totally different options than expected. The linearity of development is another assumption that is often made although developments are not necessarily linear but have many feedback loops. This can be observed especially for technological innovations (Grupp 1997).

Forecasting normally ends with the identification of the possible futures (e.g. Martino 1983, Linstone/ Thuroff 1975, Jantsch 1967, Gordon/ Helmer 1964 and many others) although some of these first approaches have already some foresight aspects

---

2   There was one case in a scenario by Shell that dealt with this option, but not a real "forecast" on the oil crisis, see also van der Heijden 1997.

in mind (see below, Linstone/ Simmonds 1977). "The technological forecasts cannot predict what will be learned by the research; they can identify research needs by identifying ranges of phenomena that will be encountered but for which knowledge is lacking. These knowledge "gaps" then provide the basis for research programs to make the improved products possible" (Martino 1983). Forecasting can therefore be needs-driven, too, and is applied to identify not only (technological) feasibilities but also needs.

Figure 1.3-2:   Many futures are possible

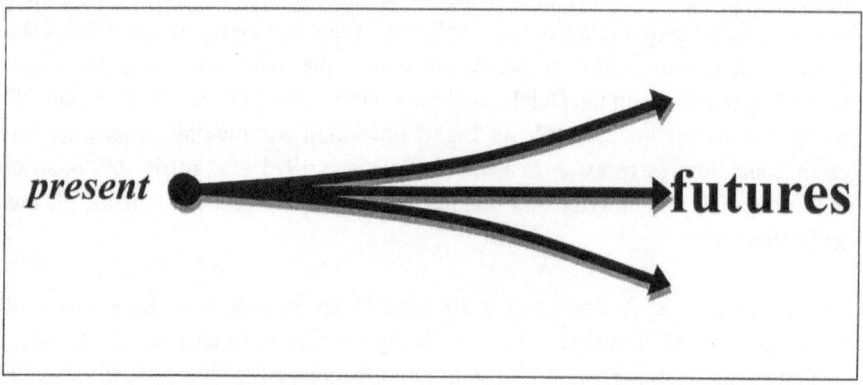

Foresight goes further. The "classical" definition is made by Ben Martin (1995 and 1996) *"foresight is the process involved in systematically attempting to look into the longer-term future of science, technology, the economy and society with the aim of identifying the areas of strategic research and the emerging of generic technologies likely to yield the greatest economic and social benefits"*.

This means not only the systematic look into the future but also the identification of strategic research. Picking the Winners, as their famous book is titled (Irvine/ Martin 1984) does not only include the look into the future but the preparation of first decisions concerning the future (fig. 1.3-3, 1.3-4 and 1.3-5 below). Foresight is not only about "picking the winners", it can also be strategically used to pick out the losers or just provide information. Foresight draws conclusions for the present and is therefore a broad-range policy instrument with sometimes various objectives (Cuhls 2000 a+b). There is not the one user and not the definite participant in foresight approaches (Cuhls 1998 and 2000 a+b). The communication effect of pre-assessing future options or decisions as well as mobilising and bringing together the different stakeholders of the innovation system ("wiring up", Martin/Johnston 1999) seems to be as important as empirical results.

The end of the 20th century has witnessed the advent of many new foresight methods and combinations thereof. Most of the experiences in organised experiments applying various foresight initiatives concerning future issues in science, technology or society were evaluated as very positive. Companies made use of the data, the media published a large number of articles, ministries reflected once more about their research priorities, and a research institution even based an evaluation on Delphi results (Cuhls/ Blind/ Grupp 1998, see section 1.6).

In most countries, the activities were supported by the research ministries or other public bodies. All foresight models try to implement communicative processes which integrate the different actors in the innovation systems ("strategic" or "distributed intelligence", see Kuhlmann et al. 1999). Most activities also attracted interest of the general public – either because of the approaching year 2000 – or because the need is felt to gain more information by looking into the future. The targets of foresight activities changed accordingly.

Therefore, foresight is conducted in order to gain more information about things to come so that today's decisions are more solidly based on available expertise than before. Foresight is more than prognosis or prediction. Implicitly, it means taking an active role in shaping the future. A possible result can be that our prognosis of today will be falsified in the future because of our new orientation and the decisions we base on foresight information. Former attempts to plan the future or to develop heuristic models of the future (in the sense of some futurological concepts) were based on the assumption that the future is pre-defined as a linear continuation of present trends (Linstone 1999 and Steinmüller 1995 give an overview; Flechtheim 1968, Helmer 1966). These approaches were not regarded as successful because they were too simplified on the one hand – and evaluated mainly by the prognosis time (e.g. if the prediction was "right" or "wrong") on the other hand. Some of the methods included different variables to match the complexity of the dynamics of the actual social, economic and technological developments, but this was also insufficient from the prognosis point of view. Nevertheless, some of these studies evoked a vivid discussion about the future (e.g. Meadows et al. 1972, Forrester 1971).

In reality, most areas of the future underlie reciprocal influences, which at present cannot be assessed. Only the visible parts, structures or framework conditions can be understood or partially influenced. If the knowledge in systems analysis theory is also taken into account, the mutual influences of systems and rules, in which the action of mankind is embedded, must also be reckoned with. An uncertainty was perceived in futures research when new experiences of the chaos theory emerged (Steinmüller 1995). The new thoughts that came up in the nineties (starting with Irvine/ Martin 1984) did not say that the future cannot be influenced directly, but made clear that the influence on future developments is strictly limited and that the impacts can only partially be estimated. Nevertheless, the future can be "prospectively monitored". The accelerating changes that a person has to adapt to socially

and psychologically make it necessary to anticipate these changes before they become reality (Helmer 1967).

Figure 1.3-3:   Selection of one of the different future options

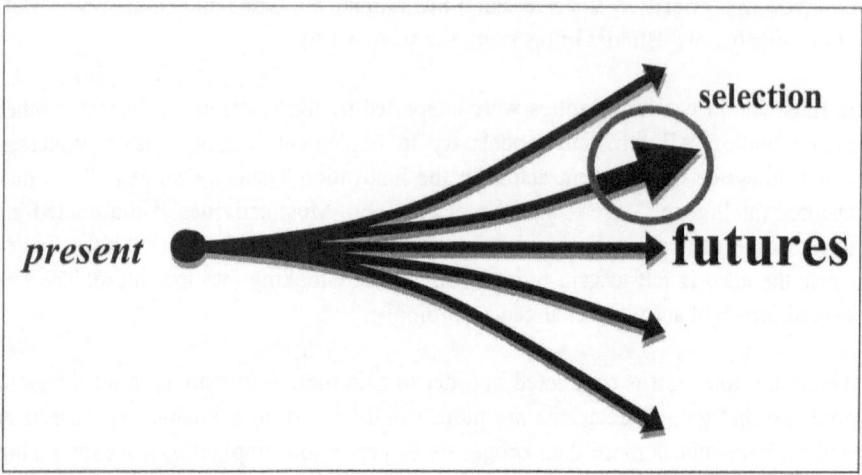

As figures 1.3-3 and 1.3-4 show, foresight starts with forecasting and the identification of *the one* or *the different* options for the future as well as their alternatives. But
foresight goes on further, to identify the most probable, the possible, wishful future
(or apply other criteria).

Figure 1.3-4:   Conclusions for the present

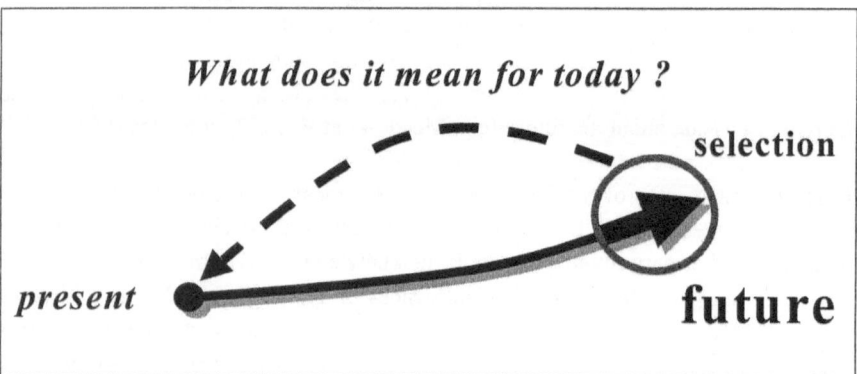

Taking these criteria into account, one option is chosen (fig. 1.3-4) to be examined
in more detail and to draw conclusions for today. If necessary, other future options
can also be examined, e.g. when there are more than one really wishful option. The

further examination can ask questions like: What does this option mean for today? Does something have to be changed? If yes, what? Who has to do it? Why? What does that evoke? Where is the change necessary? And how will the future option change when these measures are taken? If changes are unnecessary, will the option really develop as assumed before? Who is affected? What impacts does it have?

On the assumptions made, a decision has to be prepared and made (fig. 1.3-5). By this decision to opt for a certain option, the option turns into a target (fig. 1.3-6). This target can be modified but once it is set, it cannot totally be abolished without revising an explicit decision. At this point, the part foresight can play ends, and planning for the future ("forward planning"[3]) or the definite implementation of the decision starts. With foresight tools and methods the priorities can clearly be identified.

Figure 1.3-5:   Decision for one of the options

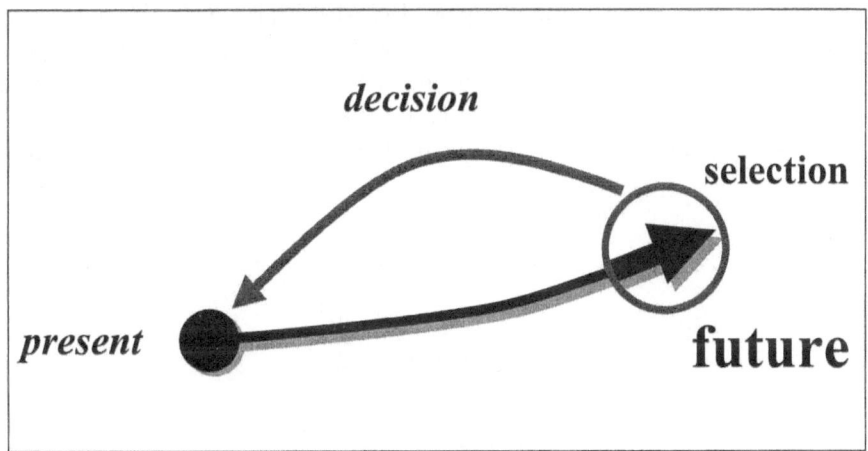

---

[3]   This expression can be found in some publications nowadays, but as planning is already future-oriented by definition, "forward" only stresses the fact that over time, the image of planning had become more and more that of static, present- or short-term-oriented planning.

Figure 1.3-6:   Defining a target

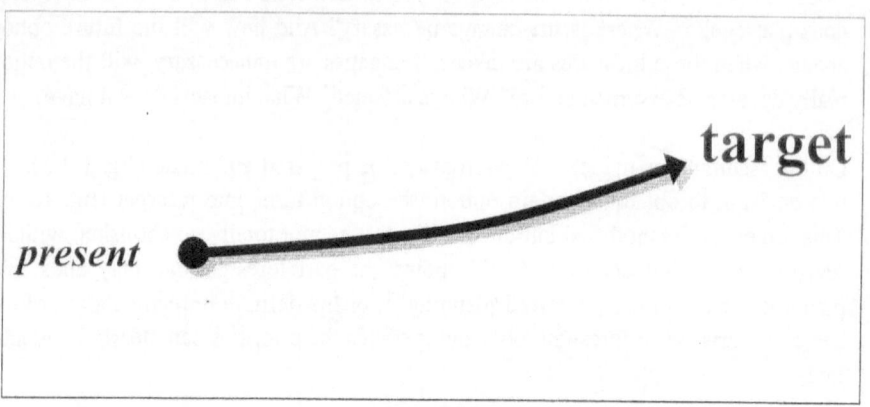

Planning: There is a link from forecasting to foresight to planning. Some parts even overlap (fig. 1.3-7). But as Coates (1985) already mentioned: "Foresight is not planning. It is merely a step in planning." But what is planning? Although everybody seems to have a feeling what it can mean, the definitions vary: "Planning is a structured process to integrate information and to conceptionalise. It fixes those factors which are anticipatively necessary to achieve a target. The result is a plan." (Bea/ Dichtl/ Schweitzer 1989). This is the definition from a business administration "textbook".

Figure 1.3-7:   Overlap of foresight, forecasting, and planning

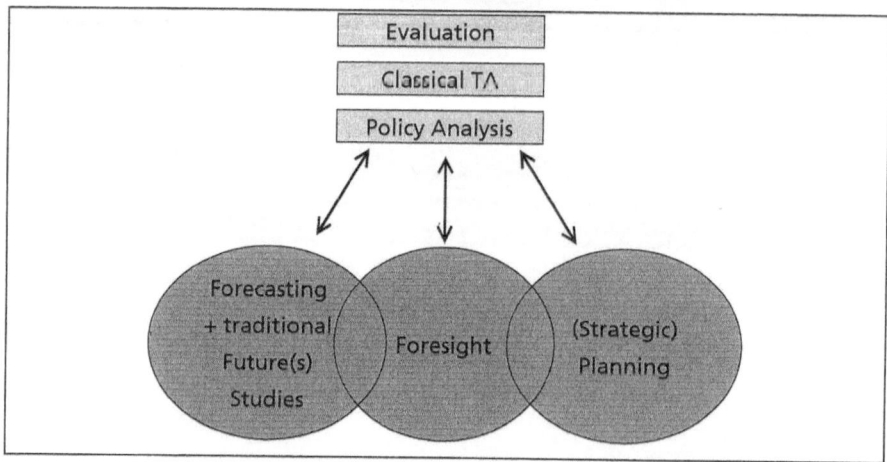

"Planning is the spiritual as well as organisationally and institutionally formed process to fix by assessments, concepts and decisions, on which ways, with which steps, in which timing and organisational succession, under which framework conditions and lastly under which costs and risks a definite goal can be reached. Anthropologically seen is planning the attempt to reduce the coincidences in the world, the diversity of possible alternatives, and the unsecurity about the future as well as the not-knowing and side-effects or feedback loop effects of actions so that minimisation of risks and a goal adequate choice of actions are possible..." (Brockhaus 1999). This is the definition from a German encyclopedia. Websters New Encyclopedic Dictionary (1993) even defines "planning" as "1. to form a plan of or for: arrange the parts or details of in advance (plan a party) or 2. to have in mind: intend". A lot of different definitions can also be observed in Mintzberg 1995.

Planning is not necessarily fixed, but once targets and the assessment of chances and risks are identified, a policy is worked out which fixes the plan to a certain extent. The strategies based on it are also the base for the long-term planning. The formal document for it is the *strategic plan* (Neske/ Wiener 1985). Planning is only possible in areas which can be directly influenced by mankind, those which can pragmatically be influenced by actions. Forecasting and foresight also try to figure out those occurencies that cannot be influenced at all or not that directly (e.g. climate change; an eclipse of the sun; the development of earth population in general, although in a region it can become close to being plannable; a vulcano eruption; earthquakes; the marketing of an innovation etc.).

Figure 1.3-8:   Planning

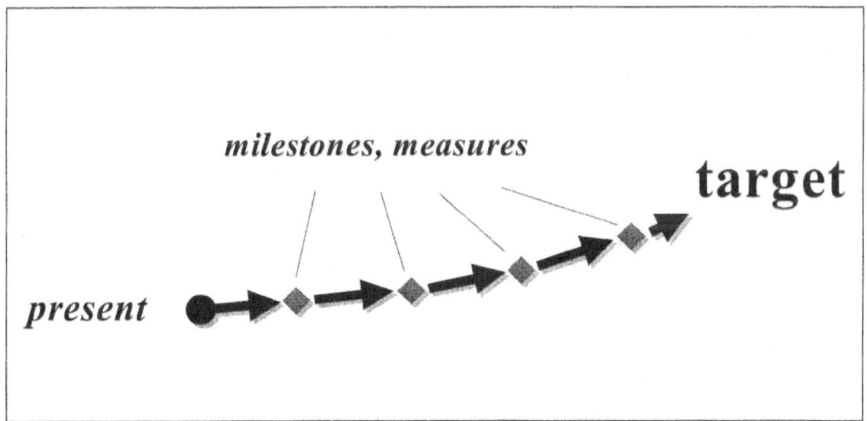

The result of planning is always something pragmatic, the *operative* plan tells what to do. The plan therefore includes milestones to reach and measures to take (fig. 1.3-8) which depend on task, target, framework conditions, and user. There are numerous tools and techniques – especially in management – to work out plans. Long-term planning and foresight often have similar perspectives. Therefore, in the preparation of the plan, foresight is one of the instruments to use. Vice versa, planning is not foresight. Often – and especially in companies – long-term planning means three to five years (Kono 1992) whereas the perspective in foresight is 10, sometimes even 20, 30 or more years. Planning is the real preparation for the things to come in the near future with a target whereas in foresight, the target as such is figured out.

1.4   Methodological tool-kit of foresight

So far foresight has not devised any clear cut methodological repertoire of its own. It tends to draw on the respective methods required from neighbouring disciplines and forms a suitable mixture of combining them. Here we mention methods from innovation economics, systems analysis, operations research, technology assessment, social and public opinion research. Although no specific, generally binding mathematical algorithms have been developed, certain approaches quite clearly appear more frequently in foresight than in other disciplines whilst other, more widely used algorithms, on the other hand, occur more rarely. In this section, we seek access to methodological aspects of foresight by asking which particular preferences are involved and how they changed over time.

Figure 1.4-1:   Foresighting methods

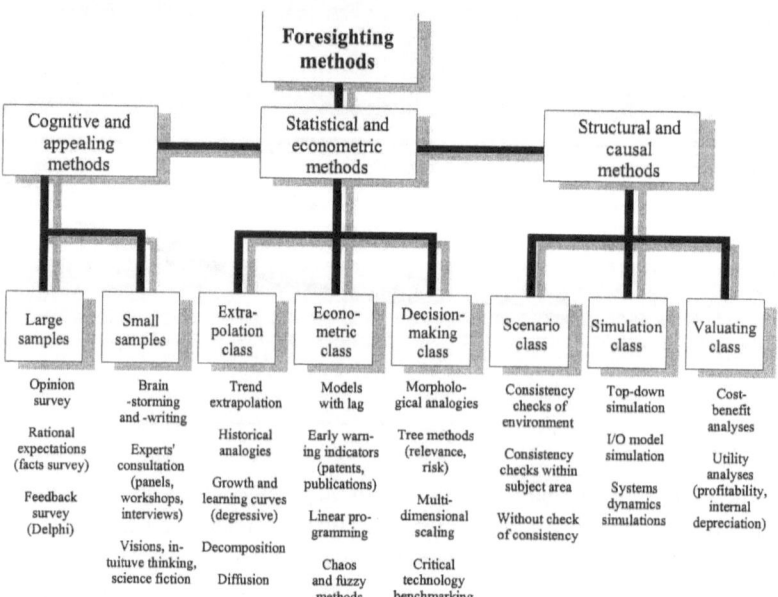

Figure 1.4-1 provides a very condensed overview of most of the more general foresight methods. They were grouped into three basic types (for earlier surveys of foresight methods see, Linstone/ Thuroff 1975 or Martino 1993). The first group we denote as "cognitive-appealing methods" and we subdivide these into two classes: large samples versus small samples. Opinion surveys in most cases require large samples in order to avoid biases. If opinions or normative topics are not asked for, but facts about the future then the nature of such surveys comes close to what economists term "rational expectation". As there are no facts on future events, respondents can only judge on their rational expectation of what might be the case in the future. If such a survey includes feedbacks, that is communication of estimations between the respondents, it is termed a Delphi survey. Delphis were most popular in forecasting exercises of the early days, but also in modern times.

The Delphi method consists of experts' judgement by means of successive iterations of a given questionnaire, to bring convergence of opinions and to identify a possible consensus (Linstone/ Thuroff 1975; Martino 1993). Each iteration constitutes a "round", and the questionnaire is the medium for the experts to interact. The Delphi method is considered especially useful for long-range aspects (20 to 30 years) as expert opinions are the only source of information available for this time horizon. The main advantage of this method is that panel members can shift position without losing face if they see convincing reasons for doing so.

The panel members will usually have widely varying estimates on each topic in the beginning of the process and do not always alter their opinion under the influence of the assessments given to them by the other panellists. It is commonly assumed that the method makes better use of group interaction (Rowe/ Wright/ Bolger 1991). If the single expert's opinion is at variance with the group ("cognitive dissonance"), this may stimulate cognitive processes which might bring information to the surface of the consciousness which is not available otherwise (Häder/ Häder 1984). The main problem of the Delphi is that the issues asked must be generated elsewhere, they do not originate from the panellists.

The literature reviewed by Jantsch (1967) distinguishes between quantitative and qualitative methods. Qualitative results always occur in the class of cognitive-appealing methods if small samples are involved. This is typically the case for brainstorming or brainwriting, for expert panels (face-to-face meetings) and individual interviews. "Intuitive thinking" is also typical for science fiction and the establishment of "visions". Brainstorming, for example, is a small group method where a relaxed period of free thinking is used to articulate ideas before more systematic evaluations take place.

The group of statistical and econometric methods we subdivide into the extrapolation class, the econometric class (in the narrow sense) and the decision-making class. Clearly, trend extrapolation is very popular as is the inference from historical analogies. Time series displaying growth or learning curves (or degressive curves) would reflect more advanced extrapolations than simple trends, as at least a rough theory of "how it goes on" is required. Further, we mention diffusion models or the decomposition of time series (shift share analysis, for instance). In the econometric class, which is of course far bigger than indicated in figure 1.4-1, especially the models which allow for time lags are useful for foresight. So-called early warning indicators such as statistics on scientific publications or patent trends which herald market results with a lag of some years can be helpful in short-term forecasting. Linear programming, chaos and fuzzy methods are among the more modern econometric tools for foresight (Nováky 2000).

Under the umbrella term "decision-making" we subsume the class of methods such as the morphological analogy, the relevance or risk trees and multi-dimensional mapping methods of all sorts. In recent years, so-called critical or key technology lists have emerged among the "relevance" methods. The relevance or tree methods consist of identifying technologies by a set of rational criteria which meet certain requirements (importance, critical nature for defence and so forth). Often, a bench-marking analysis is undertaken for comparisons of the entries in the list with one another or with other data sets. Because several criteria are usually applied, the relevance identification can become quite complicated and hence may be termed "tree". In any case, this class of methods is normative and has its foundations in systems analysis.

The trees can help to arrive at distinct levels of complexity or hierarchy and also a combination of criteria is feasible by cluster analysis or multi-dimensional scaling in order to discern simpler structures. These methods represent an attempt to reduce the analytical diversity of high-dimensional problem areas within a given low dimensional presentation space. This ultimately means that the findings on the future are derived not from statistical inference, but from graphical structures; ultimately the detection of hidden structures is the goal.

The third group of methods we term "structural-causal methods". Here we mean the scenario class, the class of simulation and the evaluating class. Scenarios can be bottom-up or top-down and they may consist in a quantitative check of consistency in the well-defined subject area or, in addition, in the "environment" to the area studied. The method consists of organising information on future possibilities and to write down alternative paths or trajectories. Scenarios will normally be composed of a mixture of quantifiable and non-quantifiable components arranged as alternative logical strings of events. They are particularly helpful if combined with other meth-ods, such as panel consultation or Delphi questionnaires. In the simulation class we mention computer simulations of all sorts but also input-output models, systems dynamics simulation and the like. If cost-benefit analysis is involved, profitability calculation is aimed at, or utility values are analysed, then we should speak of the costing or evaluating class within the causal models. If the latter methods are used for foresight, then always problems of depreciation rates of interests or return have to be solved.

Although "consultation" is regarded as a "method" (Internet page of Swedish Foresight: www.tekniskframsyn.nu), from the epistemological point of view, this is more a concept or an approach using structured interviews, panel workshops and other activities. Thus (science and) technology foresight describes a type of combined analysis and communication processes in which informed parties and stakeholders participate in a forward-looking exercise.

While trend analysis and any sort of statistical and econometric method was well suitable for the older approach of forecasting, in more recent times cognitive and appealing methods are at the forefront. Comprehensive simulation methods of all sorts were typical for the 1970s with the famous example of the report of the Club of Rome in 1971 (Meadows et al. 1972). Yet, the world models turned out to be too crude and could not resolve specific regions or problems and thus are increasingly replaced by more narrowly defined scenarios. The best way out of the difficulties is the use of "multiple perspectives" (Linstone 1999).

It is also more common now than in the past to give more attention to the feedback processes between capability and need which must link up in timely fashion for decision-making. While the traditional tools of forecasting, such as trend extrapolation, are appropriate during any stable phase but inherently fail in the chaotic states, the emphasis on communication processes increases our ability to respond capably to any anticipated situation and in particular to effective crisis management. The use of multiple perspectives is particularly helpful in this context (Linstone 1999). Foresight has changed in notion and in methods. It will probably require a much broader approach in the future than ever before. The new foresight approaches described in chapter 9 already experiment on a broader basis

1.5   Forward looking over time: the career of the Delphi method

In the previous section, we tried to give an overview over different types of methods principly available for foresight purposes. There we also made attempts to demonstrate some basic structural shifts in the use and thus the utility of the various classes of methods. Each of those methods has its special "career" in the second half of the century and it would be worthwile to comment on and to study each of those groups of methodologies. In this book, however, we limit ourselves to a more detailed analysis of the historical developments around the Delphi method. It would be worthwile, on other occasions, to consider the ups and downs of the other methods in a similar manner. It is more clearly differentiated from other surveys and termed homogeneously in the literature from various disciplines and countries.

But this book is mainly on foresight in Germany, and therefore, Delphi plays a large role. We start with hindsight and consider briefly what the function of the priestess in the old Delphi temple was, whether historical research found any impacts on politics or society in those days, and whether there was a lasting impact on the progress of mankind in prehistory. It seems to be justified to start with the Delphic oracle, as Woudenberg (1991, p. 132) reports that the name "Delphi" was intentionally coined by Kaplan, an associate professor of philosophy at the UCLA working for the RAND corporation in a research effort directed at improving the use of expert predictions in policy-making. Kaplan et al. (1950, p. 94) referred to the "principle of the oracle" as a "non-falsifyable prediction", a statement that does not have the property of being "true" or "false". Thus "Delphi" for the modern foresight method seems to be more than a simple brand name.

The foundation of the temple at Delphi and its oracle took place before recorded history. Thanks to archeologists and historians we have extensive knowledge on the functions and benefits of the oracle (Parke/ Wormell 1956). For a thousand years of recorded history the Greeks and other peoples, sometimes as private individuals, sometimes as official ambassadors, came to Delphi to consult the prophetess, who was called Pythia. Her words were taken to reveal the rules of the Gods. These prophecies were not usually intended simply to be a prediction of the future as such. Pythia's function was to tell the divine purpose in a normative way in order to shape coming events. To put it briefly and coarsely: Parke/ Wormell (1956) are convinced that the mystery of the oracle can only be accounted for on two extreme spheres or a blend of them. According to the first, the priesthood and the Pythia were deliberate charlatans who worked a traditional "hocus pocus", because it paid them well. According to this view Pythia's role was simply play-acting on her part, traditional and effective in impressing the credulous. The opposite view would be to suppose that the priesthood and the Pythia were perfectly sincere. Instead of being the deceivers, they were the deceived.

One should consider that the Delphi monastery was one of the very few spots on the earth where knowledge was accumulated, ordered and preserved. The information came in from the ambassadors through their queries and the answers were written down on metal or stone plates, several of them found by archeologists. The temple was the locus of knowledge, or, if we put it more mundanely, the Delphic oracle was probably the largest database of the ancient world. The priests could read and write; who else could do so in Greece? If due allowance is made for these circumstances, modern psychology will find no special difficulties in accounting for the operations of the Pythia and of the priests interpreting her utterances. Knowledge was intended to be used and disseminated to make the world better.

Certainly, the consultations were religious in form and not mere inquisitive speculations on the future or attempts to obtain practical shortcuts to success, but at least in earlier periods religion entered into every aspect of Greek life and there were few subjects on which the advice of Apollo was not sought (Parke/ Wormell 1956). There is no doubt that the oracle acted as an international arbitrator. It shared the rise of Hellenic civilisation to which it contributed no small part. It is no wonder that a witness of that time, Socrates, around 400 years before our time, judged: "The prophetess at Delphi (...) turned many good things towards the private and the public affairs of our country" (Socrates ca. 400 BC).

Thanks to the oracle, the Greek people learned over many generations to abstain from bloody vendetta, to apply to courts when quarrelling in private life occurred, and to solve disputes in a fair way. It can be traced back to the oracle that one should not poison the well of one's enemy and should take care of the olive trees in war. Thus the idea of the long-term oriented development of landscaping achievements we owe to the Delphic oracle. Based on this impressive historical material, let us turn now to the routes of the modern Delphi method.

In figure 1.5-1, as an illumination of the "genealogical tree" of the Delphi technique, the major steps achieved in a chronological manner are listed. The major national endeavours using the Delphi technique are taken into account, but not for example the many experimental or scientific applications where, say, 20 students are engaged in the frame of a master or doctoral thesis. Also not included are business applications on a more focussed and less sophisticated level. It has to be stressed here that we intentionally focus on large holistic surveys with a likely impact on society. For the other type of Delphi application, refer to business management text books or monographies on strategic planning where Delphi applications are mentioned among the other tools (compare Linstone/ Thuroff 1975; Martino 1993; Jantsch 1967; Cuhls 1998).

As already stated, the initial work was performed at RAND after 1948. In 1964, for the first time, a huge Delphi survey in the civil sector was published (Gordon/ Helmer 1964).

Figure 1.5-1:   Genealogical tree of Delphi

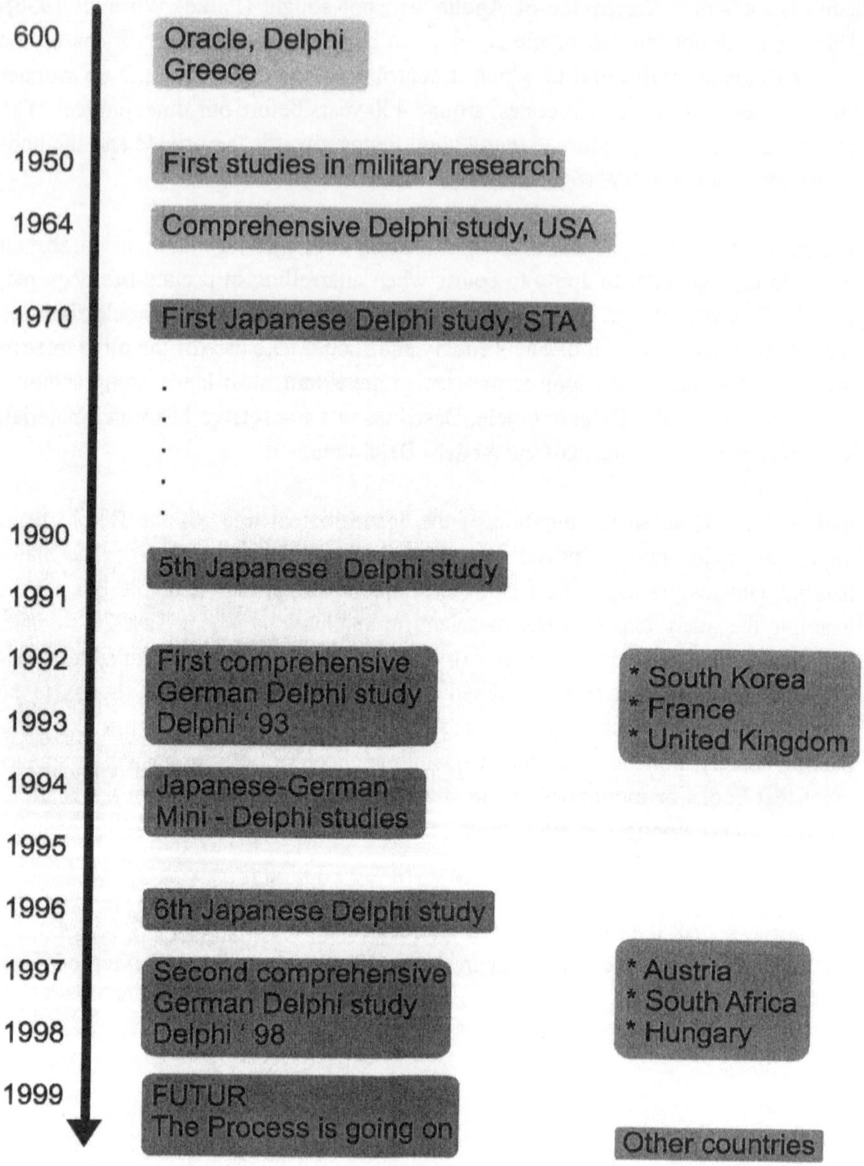

600	Oracle, Delphi Greece
1950	First studies in military research
1964	Comprehensive Delphi study, USA
1970	First Japanese Delphi study, STA
1990	
1991	5th Japanese Delphi study
1992 1993	First comprehensive German Delphi study Delphi ' 93
1994 1995	Japanese-German Mini - Delphi studies
1996	6th Japanese Delphi study
1997 1998	Second comprehensive German Delphi study Delphi ' 98
1999	FUTUR The Process is going on

* South Korea
* France
* United Kingdom

* Austria
* South Africa
* Hungary

Other countries

Shortly after this, the lead in further development and broader application of the Delphi technique was taken over by Japan. Japan started its development of S&T later than Western countries and was nevertheless immensely successful. There are many success factors for this story – and one of them was the adaptation of large foresight studies at the end of the 1960s. In Japan, the Delphi method was selected for foresight activities, and the Science and Technology Agency in 1969 started to conduct a large study on the future of science and technology. Before, in a systematic attempt, foresight knowledge from the USA was invited. The Delphi application was successful as an instrument to systematically look into the longer-term future and it did identify the areas of strategic research and the emergence of generic technologies with the potential to yield economic and social benefits (to repeat the definition). Although the first large Delphi study in Japan did not correctly describe the oil price shock and was conducted and published just before that happened, the Japanese Delphi process continued every five years. It is regarded as an update of data concerning the future. In 1997, the sixth study was finished, the seventh will be published in 2001.

Now, after more than 30 years of Japanese Delphi experiences, is it possible to draw a retrospective? Cuhls (1998), with degrees in Japanese Studies and Business Administration, in her doctoral thesis dealt with exactly this issue, and by translating the 1971 Japanese Delphi from the Japanese into German, provides the following answer. Japanese S&T policies are less consistent than is commonly believed and represent an assortment of policy measures, pragmatically devised to address the diverse, ever-changing and sometimes conflicting needs for a broad range of issues. Foresight in Japan brought in (and still brings in) elements to moderate or negotiate between the social interest groups. Foresight results provide the code to communicate between social actors in science, technology and society.

Foresight was developed and used without such sociological reflection. To cite two examples Cuhls investigated among many more (Cuhls 1998 and 2000), the foresight topics on early earthquake detection were always corrected in subsequent studies towards the more distant future. Yet, even before the Kôbe disaster, which was of course completely unforeseen, the foresighting activities kept this issue on the agenda in S&T, admittedly on a low level. Every five years, an earthquake prognosis plan is elaborated and the Ministry of International Trade and Industry (MITI) finances geological surveys. In 1993, before the Kôbe earthquake, a special measures law for earthquake disaster prevention passed the Diet (the Japanese parliament). The Japanese Delphi studies and the ever shifting realisation times were helpful in pointing to the unresolved issues in years of no earthquakes with little public and policy awareness of these issues (see also chapter 6).

An other example relates to solar cells. R&D for solar cells has been at the top of Japanese priorities for years. The earlier forecasts were delayed in each subsequent survey, but since the mid-1980s they are stable. Clear impacts of the Delphi studies on government priority programmes are visible. Some firms overdid their R&D investment because they were as optimistic as the early forecasters, but the MITI backed this over-investment and thus accelerated the real progress – a clear case of self-fulfilling prophecy. Also diverse regulation changes helped to get the mass production of solar cells started. As a matter of fact, foresight provided the tools for S&T policy in Japan required by modern theories.

For the Japanese policy, it was especially interesting to answer the following question (and this question is also asked by other governments, too, now): How should we proceed with the long-term application-oriented basic research of the hyphenated type? This extension is no mistake, we really mean *long-term application-oriented basic research*. This is the research where one does not know what will be found out in the laboratory in the next month or year, but it is research which does not only satisfy scientific curiosity and the enhancement of knowledge. It is research with a definite long-term economic or social perspective. Let me mention climate research, health research, environmental research and so forth. In days of low budgets many business and policy-makers think it is impossible to support each piece of interesting research only for the sake of good quality. One has to discuss the long-term orientation in which we invested our precious money. The public is convinced that science and technology are partly responsible for modern bottlenecks and problems and hence has a right to learn about priorities in technology and also the opposite, the non-priorities, what is down at the end of the list of priorities.

Consider the situation in which a company or a ministry has to decide which of two research programmes to support, A or B. Programme A is proposed from faculty A and industry A and the peers from discipline A have given their reviews. Programme B in conjunction with industry B originates from faculty B and the peers of discipline B made up their minds. Everybody did her or his best. But how to decide between them? Do the peers know each other? Our science and technology system of tomorrow needs, alongside with disciplinary peers, new instruments to mediate between A and B, and here is another function of foresight, across the board.

A second argument here is that they all have their stakes in the matter. They come from the technology provider side. But do they really know what is needed?

Most sociologists of science assume that there is a positive relationship between involvement in a research area and assessments of it and that this relationship derives from the tendency of scientists to select problems in areas where there is high pay-off for successful solutions and career. The tendency to overrate fields in which a person works may be termed "bias". Not only a tendency toward positive bias for fields in which researchers have beeen active was found, but also that this bias is stronger in less innovative sub-fields. As market signals fail to be useful for business strategy in the long run and expert assessment is not always objective, Delphi surveys may play a part in science and innovation management.

There are two examples from the first German Delphi '93: specialist experts and thus future knowledge may not be available in some countries. The availability of experts in the case of biotechnology in Germany was mixed. Among the 73 respondents who were all experts in biotechnology, many did not answer in particular subareas (most expressed for tissue and organs). The largest number of specialist experts (i.e. those working in the sub-area) among all experts in Germany is found in molecular biology, but not in the sub-area of tissue and organs. An almost perfect correlation was found between the number of experts and their rating of German research performance. In sub-areas where we know more, we are good. In sub-areas where we are not advanced, we know little of the opportunities.

A test for Delphi expert bias in the energy area tends to support this view. Top experts rate the importance of their own research speciality significantly higher than the other experts - both in Japan and in Germany. At the same time, the top experts downplay technical constraints in Germany (less so in Japan) in their own working area (see Cuhls/ Kuwahara 1994).

In the Delphi '98, this is not so obvious. There are topics for which the specialist knowledge experts see more problems (or ask for more measures to be taken), but for others all other persons ask for more measures. In some cases, the special experts rate the topic to become reality earlier than the "medium" and "lower level" experts, in other cases, they are much more reluctant with a prognosis on the time horizon. What can be observed is that in the first round, more experts claimed to work on the field (13.5 %) than in the second round (10.18 %). This can have several reasons (see chapter 3).

With the resurrection of foresight in general and the possibilities to filter all these "options" of different actors, the Delphi technique was taken out of the toolbox and implemented in Europe in a different manner than in the early years. In the new wave of large-scale government foresight in Europe, Dutch and German government agencies and similar bodies were among the first, with France and the United Kingdom joining in quickly. The Germans organised a learning phase starting both from the "mediating" publication of Martin and Irvine (1984) as well as from Japanese experiences and co-operated in their first Delphi with the Japanese fifth endeavour (Cuhls/ Kuwahara 1994). France in turn followed in just copying the German approach. In none of these countries was a sole resort to the Delphi technique considered useful. In the Netherlands, Delphi methods were not embarked upon at all, whereas in Germany parallel approaches are reported, some using the Delphi method, others not. The same is true for France where a Delphi survey and the critical technologies approach (see figure 1.5-1 or Grupp 1999) were pursued in parallel and organised by different, even competing ministries. Again in co-operation between Japanese and German institutions, joint methodological developments were achieved in the frame of a "Mini-Delphi".

Although in the United Kingdom a first national foresight programme was set up as the central activity, a programme which heavily relied on panel work and related methods, a Delphi survey was added to it in the sense of a supportive tool (at least in the first programme and contributed in the last minute to some of the panel reports). In the Republic of Korea, Austria, Hungary, South Africa and in developing countries, the Delphi technique seems to be a central backbone in government foresight due to its comprehensiveness.

The Delphi technique as a foresight tool seems to possess certain degrees of invariance to survive in the changing challenges of the past 50 years. The method could serve different understandings of forecasting or foresight and was probably understood by the users as being relevant for covering technical perspectives, organizational perspectives, but also personal perspectives. The individual could express a distinctly different opinion as compared to the group perspective and this to a differing degree between the technical details under scrutiny. As multiple perspectives are recommended for decision-making, (Linstone 1997; Linstone/ Mitroff 1994; Linstone 1998) the Delphi technique seems to have appeal in quite diverse situations which touch the long-range scales. As it can be shown in controlled scientific experiments that the position of Delphi estimates is not better than those of other consensus-oriented methods (Woudenberg 1991) it must be the communicative force of Delphi approaches that facilitates the switching between different perspectives. What users especially like are the sets of data about the future that are gathered. Writing down future topics seems to have an immense psychological effect because it transfers implicit to tacit knowledge to the more visible, explicit, and therefore transferable knowledge.

Nevertheless, the danger that many persons regard this as "the future" that "will come true" cannot be neglected. When the media in Germany used Delphi '98 data for an outlook into the next century, they often made the mistake of arguing that the future will be like it is described in Delphi '98 disregarding that the decisions of today (or non-decisions) have a strong effect on the things to come and that Delphi can only provide "potential answers" to problems that can already be identified today.

The second problematic point remains the interface to implementation. As the providers of foresight results and the users, which means the decision-makers, are in most cases not the same persons, there remain the difficulties

(1) of bringing them together

(2) of linking the needs of the users and the concepts of the methodologies very early

(3) of making potential users aware of the possibilities (marketing) so that they have the choice

(4)   of establishing mechanisms of transfer

(5)   of delivering results that are useful

(6)   of involving persons who have the power to decide and implement.

Until now, the use of foresight results in Germany was based on ad hoc activities (see below). They were very useful and there were a lot of them, but a more strategic approach would certainly bring more results although this strategy will have its limits, too, as can be seen from the chapters below.

1.6   Benefits of foresight in general

Several lessons can be learned from the application of foresight methods. Firstly, it is important to note that a foresight activity should not be a single event but should rather become part of a broader strategy which deals with strategic orientation. Secondly, the individual results of a survey should trigger various follow-up activities within the organisation, for example, workshops on selected items. Thirdly, going through the process of a foresight survey itself is a very valuable undertaking, since great numbers of experts are motivated to think critically about future scenarios favoured or rejected by their peer colleagues. Fourthly, for a company, the benefits of a foresight survey should not only be seen as gains in information and reputation among its clients, but also extended to the internal situation: the strategies for dealing with challenges of the future must become broad company issues which are to be discussed and supported by many employees, thereby contributing to an increase of in-house motivation and identification.

There are different actors who can benefit differently from foresight activities. Details from the Delphi '98 are explained in chapter 3.

From the social point of view, the direction to be taken in the future may be derived from the increasing demands made on technological development in terms of minimal use of resources, elimination of emissions, recycling economy and sustainable development. These demands require the creation of the new framework conditions, especially those of a non-technical nature, such as legal regulations. Equally important as such ecological problems is the socio-political dimension - in particular the unemployment problem. From the point of view of technology policy, a form of technological development is needed which encourages wide-ranging participation by employees in various sectors, and firms of varying size, which leads to an open market with no specific centralised structure.

It is in the nature of long-term foresight that it is burdened with a high degree of uncertainty as to how the decision-making groups will behave; it is not unusual for wishful thinking, arising from the most diverse motives, to be presented as a probable future event. Taking the long-term view, the motivating power of guiding visions is helpful in that it releases social energies and the willingness to undertake concerted action. Long-term lead projects in technology can produce lasting motivation and unite powers which can work towards problem-solving requirements recognisable in the long term, and also produce successes along the way (through desirable multiplier effects).

Lead projects in technology which represent outline solutions to large, global, economic, social and ecological problems, and especially the visionary view of technological development and the challenges now facing us, throw up other, more radical questions of technology policy than those set out here for the time being. It has been possible to indicate that technology assessment through foresight can itself provide the key to far-reaching changes in future policy. The technology policy of tomorrow must be in place to shape technology policy in the long run.

In Germany, generally, there is a public tendency to be critical about new technology, often without going into any detail. After some foresight studies were published - rich in presenting visions of detailed trends in science and technology - several "second thought" articles concerning the public understanding of technology by science journalists were published (for a list see Cuhls 1998). The message in these articles is basically that dogmatic scepticism against new technology as such should be replaced by public reservations against *certain* technologies. A *technology-specific* public debate on the future of the so-called "science and technology nation" need to be triggered off. From these observations one is tempted to conclude that the assessment and foresight processes have a lasting and direct impact on society as it affects our notions of future technology. By reflecting future opportunities and impacts of technology, we reflect our procedures to get there.

2 Outline of the Second Comprehensive Outline of Delphi Study in Germany

2.1 What are the key questions?

The German economy is proud of its high export quotas. The German market is open to international competitors, and Germany itself is at the centre of a far-reaching innovation competition. However, many problem areas remain and make stringent requirements on the economy: setting priorities, the allocation of financial resources, and the strategic orientation of research and development in Germany are all under pressure.

Without an effective assessment of the most important areas in which innovation takes place and the requirements for success based on a worldwide comparison, research and development in Germany cannot develop the necessary degree of effectiveness to make innovative leaps.

For this reason, in 1995, the Federal Ministry for Education and Research (BMBF) took the initiative and resolved to finance and carry out a *foresight activity*. The activity was in line with previous studies like the Japanese – German Delphi study published 1993 (Delphi I or Delphi '93) and the Mini-Delphi studies which were supposed to develop the methodology further.

The Fraunhofer Institute for Systems and Innovation Research (ISI) was given the task of managing this project. Federal Minister Dr. Jürgen Rüttgers established a *steering committee* made up of prominent members from science, industry, and the media. The committee was given the task of advising the Ministry in all decisions concerning the establishment of important framework guidelines.

Steering Committee:
Chairman: Prof. Dr. Gerhard Zeidler, DEKRA
Vice-Chairman: Prof. Dr. Hans Jürgen Quadbeck-Seeger, BASF
Reiner Korbmann, Bild der Wissenschaft
Dr. Wilhelm Krull, Volkswagen-Stiftung
Prof. Dr. Friedhelm Neidhardt, Wissenschaftszentrum Berlin für Sozialforschung
Dr. Eberhard Rauch, Bayerische Vereinsbank
Dr. Tom Sommerlatte, Arthur D. Little
Prof. Dr. Frank Steglich, TH Darmstadt and
Prof. Dr. Cornelius Weiss, Universität Leipzig

The goal was to provide answers to the following, critical *key questions* - and perhaps to other questions that had not previously arisen:

- In which areas of innovation can significant advances be expected to take place during the next 30 years?
- What success concepts are linked to these?
- What impact can these significant advances be expected to have on economic development?
- In particular, what impact can they be expected to have on work and employment?
- How can technological innovation contribute to the solution of ecological problems?
- How will the development of society be affected by advances in innovation?
- Which results of research and development will produce the greatest increase in human knowledge?
- Within what time periods can the success concepts in the individual areas under study be realised?
- Which countries currently exhibit the highest degree of advancement in the various areas of research and development?
- What steps will be required to permit Germany to keep pace or even become a leader in those areas of R&D in which it is currently perceived as being weak, and how can this be translated into practical success?
- What problems can be expected to arise if the anticipated innovations are realised and utilised, and the resulting products must, at some point in the future, be disposed of?

The steering committee pursued these key questions concerning the future. They were translated into the criteria of the Delphi questionnaire (see 2.3). As an initial step, the steering committee defined the most important areas of innovation in the future. They functioned as the "headings" of the Delphi questionnaires and started with the following questions:

What areas of innovation exhibit the dynamics with the greatest significance to the economy and society? What themes of the future can be influenced by science and technology? The developments exhibiting the greatest dynamics can be grouped into 12 different areas, some of which overlap.

1. **Information and Communication**
2. **Service and Consumption**
3. **Production and Management**
4. **Chemistry and Materials**
5. **Health and Life Processes**
6. **Agriculture and Food**
7. **Environment and Nature**
8. **Energy and Resources**
9. **Construction and Living**
10. **Mobility and Transport**
11. **Space**
12. **Big Science Experiments**

A total of 1,070 future "predictions" in the form of hypotheses were prepared on these fields – a list that, even without answers, is important and interesting. That is why the individual theses were reproduced in detail in the methodology and data volume of the German report.

2.2 Organisation of the Delphi process

The Delphi process began with the establishment of the steering committee. On its own, the nine-member committee would be unable to deal with an all-inclusive summary of innovative ideas. Therefore, more than 100 individuals with specialised knowledge were called upon from such diverse areas as industry, higher education, and other institutions. The persons were selected according to different criteria and appointed by BMBF. They were responsible for gathering the most important theses on the above-mentioned fields from the area of research and development.

Based on good experiences in the past, a portion of the developed theses were also worked out in co-operation with the Japanese National Institute of Science and Technology Policy (NISTEP in Tokyo), which was, at the time, organising the sixth Japanese study on the future of science and technology. This provided a means of performing international comparisons. This also provided a means of determining whether additional surprises could be expected to come out of Asia, or whether

German "blinkers" were preventing us from having an objective view of the future. In addition, care was taken to ensure that an additional portion of the theses were relevant to the first, German Delphi survey in order to permit time-line comparisons. Has our assessment altered in the last five years?

Figure 2.2-1: Organisation of the Delphi process

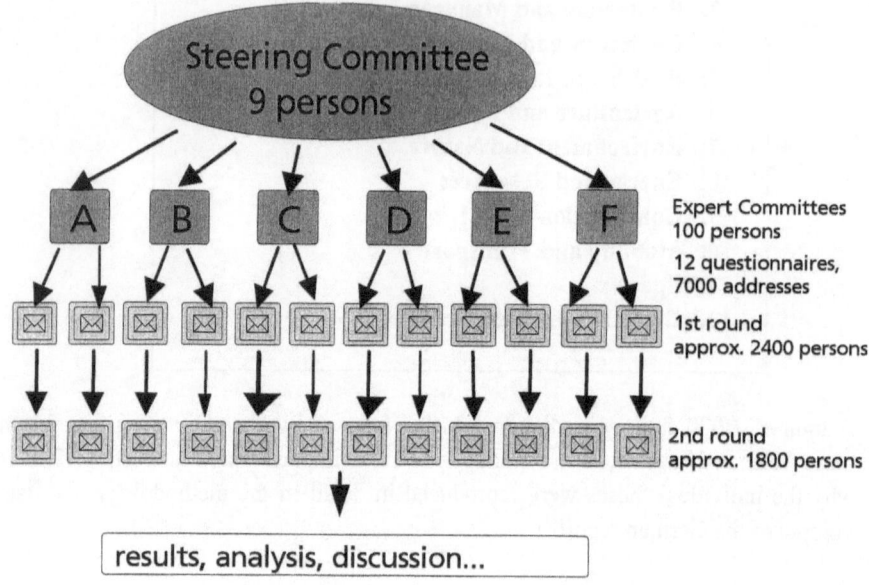

All remaining assessments were undertaken by a significantly larger circle of specialists in the various areas of science and development. The reason for this lies in the fact that, with a Delphi survey, the larger the number of participants, the greater the precision of the expected results. The study procedure involved the application of the Delphi methodology, that has been extensively tested by expert groups for these types of evaluations of the future. Since the future cannot be ascertained in advance, but a group is in a position to know or to anticipate more than a single individual, the Delphi methodology calls upon the experiential knowledge of experts.

Two, so-called "rounds" of surveys form the core of the Delphi process. The theses prepared by the specialist commissions are presented, in the form of a questionnaire, to a large number of experts for their assessment. The responses are then evaluated and returned to the same group of individuals. During this second round, the experts are asked to rethink their responses in the light of the assessments made by their colleagues, thus offering them an opportunity to alter their opinions. Since anonymity is maintained, no one loses face or must justify themselves if they change their mind.

In some Delphi studies, already the preparation and formulation of the theses is called "round" (Kagaku Gijutsuchô Keikakukyoku 1971). In the German study, we did not call it "round" but have to admit that this is the crucial phase of a Delphi study. Without proper preparation and carefully considered formulation, a study like this cannot be successful. In surveys, only answers for the questions posed can be received. More important in Delphi studies: only those topics that are formulated and the way they are formulated can be assessed. Therefore, in the Delphi '98, the preparation phase went through different stages (fig. 2.2-2).

Figure 2.2-2:   Preparation phase of Delphi '98

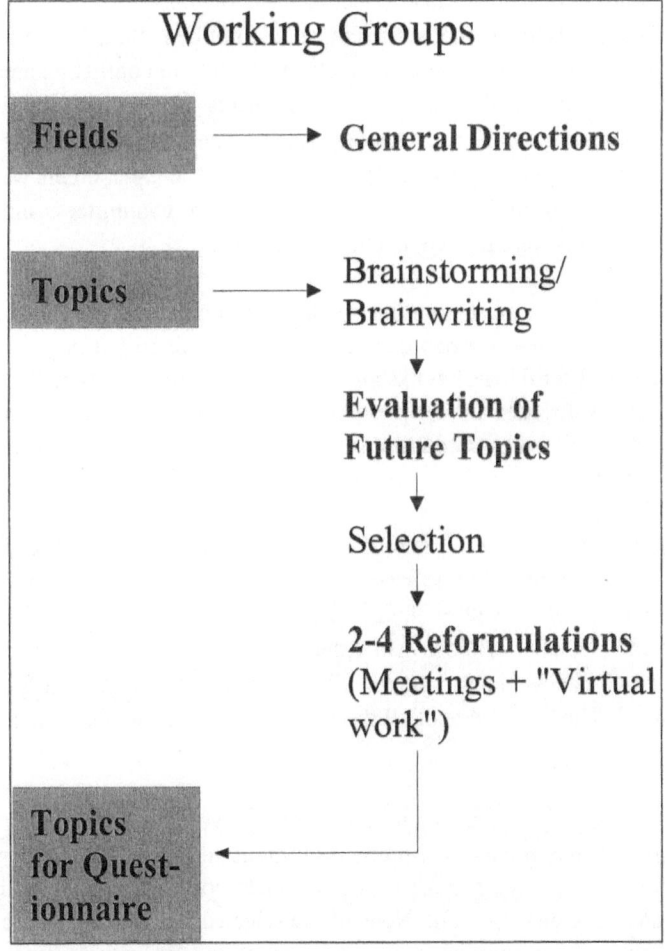

The subject areas formulated by the steering committee were structured in broader future directions as a first step. Then a brainstorming or brainwriting followed, in which topics should be formulated that

- were regarded as important developments of the future
- were feasible and realisable during the next 30 years
- meet a specific need of society and
- are not Utopian or wishful thinking.

Topics from Delphi '93 and the Japanese sixth Delphi that was prepared in parallel were added to this list. The whole list was mixed, so that the experts in the committee did not know where the topic stemmed from.

The second step was a kind of questionnaire in which the topics were written. The experts had to give "marks" (like in German schools ranking from "1" (very good, should be included) to "5" (not sufficient, should not be included or re-formulated). The assessments in this step were relatively clear. Only in conflicting areas (like the field Energy), both 1 and 5 occured as assessments for one single topic. In these cases, the topic was kept for further discussions. The next step was re-formulation of topics. Similar topics were sometimes included in one topic, others were so complex that they had to be divided. In some expert committees, only two re-formulation rounds were needed, in others up to five.

The last step was – in most committees – a workshop for the last adjustment. Most topic lists were still too long so that some topics had to be left out. In some committees, the need for this last workshop was not felt, so it was cancelled. The list of topics was then included in the questionnaires. Parallel to these activities, the experts who were to answer the questionnaires were identified. But who is an expert – and for what?

The definition of exactly who is considered to be an expert is very broad. The individuals surveyed included those who are themselves actively carrying out research in a particular field, as well as those who regularly obtain first-hand information about the field. A total of more than 7,000 individuals were approached, of whom more than 2,400 actively participated, with nearly 2,000 remaining at the end of the second round. The knowledge of all these individuals was mobilised and communicated.

The surveyed experts come from backgrounds as diverse as industry, higher education, public service, private non-profit institutions (e.g. the Fraunhofer Society or the Max Planck Society), and associations. In addition, they should be involved in research and development work. None of the selected individuals will be in a position to answer every question in the entire questionnaire with "a great degree of specialised knowledge", as this would mean he or she were active in every area covered. For this reason, individuals who regularly read relevant scientific publications, are in contact with the various researchers in the field, or were themselves active in the field in the past were also included. In other words, those are the individuals having "average" or even "little" expert knowledge about individual theses. This provides a means of making the results more relevant, should one or more of the specialists in a given field have an extremely strongly held opinion. The results from the specialists can be studied individually in the German data volume.

A multiple choice type questionnaire was used, that required the experts to respond to the theses by simply checking the box that most closely reflected their opinion. This, however does present a dilemma when performing a written survey of a large number of individuals: opinions about complicated topics must be reduced to simple responses. Large comment areas were, however, provided, should the individual wish to provide answers in greater depth.

The initial result of the survey was a large volume of data. The responses serve to provide an intimation of future developments, thus allowing a structured communication process about the future to be established. The fact that some areas of the future are already being contemplated today gains time to slow down or halt evidently false developments, or to start up or accelerate necessary innovations. Thus, Delphi studies provide no immutable picture of the future, but instead, offer a foundation of information for the decision of what is to be done – or not done - today. How the future will actually develop depends on the decisions made today. Therefore, the actual development can differ greatly from today's assessments.

What actually takes place in the individual areas? In order to provide readers with an overview, the chapters 3 and 4 will introduce the major directions of the future (the most significant results), followed by an introduction to the twelve topic areas looked at in the study. How do the majority of the experts assess the individual trends? In which areas are German research or industry "strong"? Where do opportunities lie – where the risks? What, in the opinion of the surveyed experts, must be done?

2.3   Questionnaire

The questionnaires prepared included topics which had to be assessed by marking with a cross. Figure 2.3-1 shows an extract from one of the questionnaires.

Figure 2.3-1:   Outline of a questionnaire (partial)

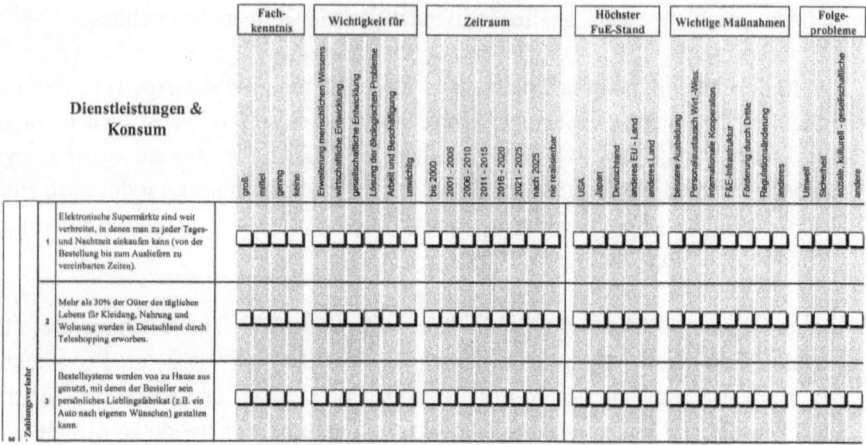

The wordings of the topics contain stereotype words which distinguish between scientific, technical or economic progress:

"Clarification" ... Principles are clarified from a theoretical viewpoint or phenomena discovered in the context of scientific research.

"Development" ... Work is carried out on implementation on a technical level, which means, for example, that the first model (e.g. a prototype) is built.

"Practical use" ... Applicability is verified in economic practice. The topic leads to practical applications, which means that the first object to be marketed is built.

"Widespread use" ... Objects used in practice are built in large numbers and under economic conditions.

An initial assessment was then to be made of what specialist knowledge the expert being questioned possesses, in order to be able to find out later whether the experts with extensive specialist knowledge assess their subject differently than people with less specialist knowledge. Secondly, questions were asked according to the *importance of the contributions to problem-solving* in terms of

• the enlargement of human knowledge,

• economic development,

• social development,

- solving ecological problems,

- work and employment.

The third point to be assessed was to be the possible *time of realisation*, in order to establish whether it is a long or shorter period to which the further appraisal refers, and to establish the chronological framework for the scope for action. The time horizon was subdivided into five-year intervals, supplemented by the period until the year 2000 and an open-ended interval after 2025.

The next question referred to the *current position with regard to research and development*. Which country is viewed as the leader, the USA, Japan, Germany, another country in the EU, or another country outside the EU?

In addition, a question was asked about how the *general economic background* is assessed and which *measures* are appropriate for its improvement:

better training or education and consolidation of the qualifications of researchers, technicians and research assistants;

exchange between science and industry in the area of personnel and knowledge within Germany, e.g. between universities and companies;

international co-operation on an operating and project level or for the reciprocal exchange of knowledge or personnel;

improvement in the research infrastructure, e.g. establishment of institutes, databases and provision of venture capital;

support by a third party (state, foundations etc.), e.g. more state financing of key projects, intangible measures or similar;

change in regulations: this can cover both deregulation, the strengthening of existing regulations, re-regulation or other changes in the state-determined general framework (laws, standards, ordinances, technical instructions, charges etc.).

Finally, a question was asked about *possible follow-up problems*: Do these lie in the *environmental area?* Can *safety problems* arise? Are the resultant problems to be expected more in the *social or the socio-cultural* area? Or are completely different problems emerging to confront us?

Alternative possible solutions could also be mentioned. It was expected that some interesting new topics could be mentioned which would be included in the second round questionnaire. But although very interesting comments were made, only very few alternative solutions ready to be included in the second round were mentioned. Particularly instructive are the comments, which range from spontaneous exclamations to detailed treatises on the subject. The latter offer a rich source of information for analysis.

The final part of the questionnaire consisted - only in the first survey round - of so-called "megatrends". These were brought up for discussion to find out which global developments technology experts anticipate and whether they expect these trends to have an influence on the development of science and technology (see chapter 5).

2.4   Who was surveyed?

So who were the people questioned in the study? Who is an expert on the future? As no-one can know exactly how the future will turn out, as many people as possible ought to take part in a survey concerning the future. For it has been proved that with a large number of responses individual errors of estimation can be evened out and that consequently the likelihood of an "accurate prognosis" is greater. However, those questioned had to be able to examine the topics closely, in order to be able to give any judgement at all. Therefore an attempt was made to build up an address database which includes groups of people who know a lot about the different fields. As far as possible, a third of these people were intended to come from industry, a third from universities and a third from other research institutes, the civil service and associations.

Table 2.4-1:   Number of questionnaires received by subject area

Delphi '98 Fields	Answers	
	1st Round	2nd Round
1. Information and Communication	287	206
2. Services and Consumption	215	163
3. Management and Production	229	179
4. Chemistry and Materials	260	206
5. Health and Life Sciences	255	182
6. Agriculture and Food	206	140
7. Environment and Nature	282	209
8. Energy and Resources	246	187
9. Building and Living	110	94
10. Mobility and Transport	150	122
11. Space	97	77
12. Big Science Experiments	116	91
All	2.453	1.856

In order to receive around 100 questionnaires capable of being analysed per subject area and, as far as possible, at least 100 responses per question, at first approx.

7,000 people were contacted. An offer was made to them to request a further questionnaire if they also felt qualified in another field, or to exchange their questionnaire for another. A total of over 2,000 people filled in 2,453 questionnaires. This is a gratifying result, as an analysable return of more than 30 % (in some subject areas of Delphi '98 it is actually almost 50 %) exceeds the experience figures in Germany. In some areas, e.g. Chemistry and Materials, Health and Life Processes, Energy and Resources and Information and Communication, well over 200 responses were received. In fields such as Big Science Experiments or Space on the other hand, we could not expect as many responses, as in Germany the number of researchers in these fields is lower than in others. Shown in the following table (2.4-1) is the number of questionnaires received in the first round and also the number of experts participating in the second round. Those people who replied anonymously were unfortunately no longer able to take part in the second round. The returns at the end of the second round are in the right-hand column and amount to approx. 75 % of the questionnaires received in the first round.

The distribution of a third of responses across industry, universities and other research institutes, associations and the civil service was, as figure 2.4-1 shows, largely successful, as in the first German Delphi project. Most of these people work in research and development.

Figure 2.4-1:   Employers of the experts questioned in Delphi '93 and Delphi '98

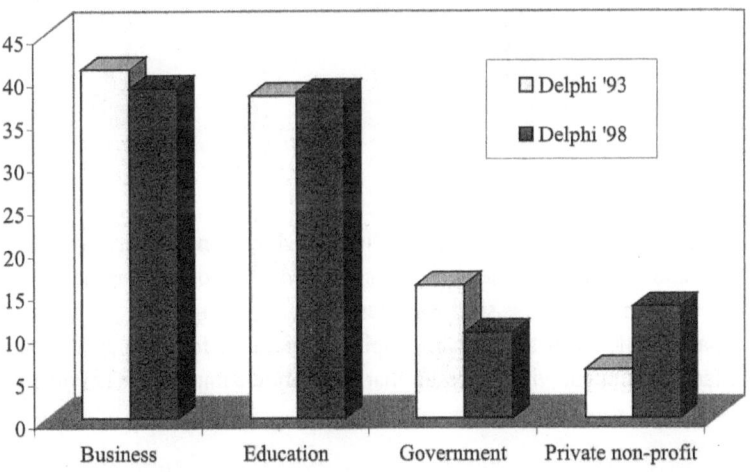

Unfortunately, only just over 5 % of women took part. This is only slightly more than in the first Delphi study, but represents roughly the proportion of women in research and development. The age distribution of the experts shows a maximum in the 50 to 60 year old age range and an approximately even distribution among 30 to

40 year olds, 40 to 50 year olds and 60 to 70 year olds. Individuals over 70 years of age and those younger than 30 are hardly represented. The latter had certainly not (yet) been classified as experts in the databases and no questionnaires had therefore been sent to them. As the individuals questioned were asked to fill in the question-naire themselves if possible, not many of the (mostly older) superiors must have passed on the questionnaire to their staff. It would therefore be interesting at some time to question a control group of younger people and one consisting solely of women, to ascertain whether they assess the future differently and would set other research priorities.

Figure 2.4-2:   Age distribution of the Delphi '98 experts in comparison with Delphi '93

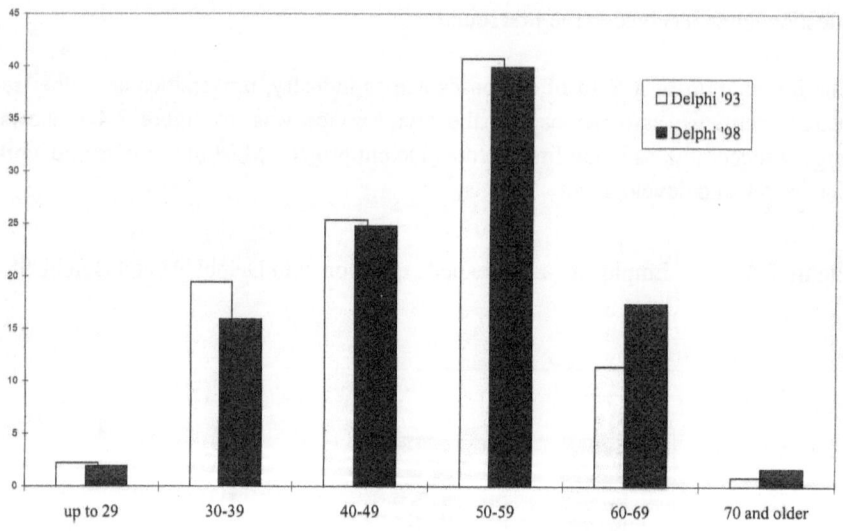

The data of the experts questioned is distributed in a similar way to the first German Delphi and the sixth Japanese Delphi study. For 323 topics, comparisons are possible in principle with the sixth Japanese study (see chapter 7). 113 topics can be compared with the first German Delphi study and a further 200 theses have been similarly continued, with the result that qualitative comparisons become possible.

2.5 Method of calculation and formulae used

In order to be able to understand the graphical and tabular illustrations in the following chapters, it is important to visualise the bases on which the data are calculated. In this chapter, therefore, the methods of calculation and their associated formulae for the separate categories are shown. The basic calculations, explained below, serve as the basis of the diagrams and tables used in this book. The descriptions of the results from the first and the second round of the technical main section of the survey are identical, so that the results tables can also be understood on this basis where the data are presented (e.g. the data diskette or the German data volume). In the German volume and the data base on internet (www.isi.fhg.de) or diskette, in the first line (1) the results of the first round are quoted, in the second line (2) those of the second round and in the third line (FK1) the assessments of the experts with the greatest specialist knowledge. Therefore, the user of the data can check and choose himself, which data set is the relevant one for him or her.

The first column, which was to be marked with a cross in the questionnaire, concerned the self-assessment of the *specialist knowledge* as "very high" (designated by Fk1), "medium" (Fk2), "low" (Fk3) and "no background in the subject" (Fk4). By virtue of this self-assessment, the *present state of knowledge* based on the respondents was calculated for each topic. People who classified themselves as having "no background in the subject" were not included in the overall calculation as they did not have to continue answering the questions.

In the graphics on the present state of knowledge, the absolute numbers of respondents, namely n_{Fk1}, n_{Fk2} and n_{Fk3}, are considered. The absolute number of respondents per topic (A^*) is therefore calculated as follows:

$$A^* = n_{Fk1} + n_{Fk2} + n_{Fk3}$$

In the individual columns of the results tables on specialist knowledge, percentage figures are given. These refer to the absolute number of respondents per topic ($A^* = 100\%$). Consequently, the results are:

$$Fk1 = n_{Fk1} \times 100 / A^*$$
$$Fk2 = n_{Fk2} \times 100 / A^*$$
$$Fk3 = n_{Fk3} \times 100 / A^*$$

In the individual columns of the results tables on *importance for*, percentage figures are given. These refer to the absolute number n_{Wi} of respondents per topic ($A^* = 100\%$). Consequently, the result for the importance is:

$$\text{Importance for ...} = n_{Wi} \times 100 / A^*$$

The absolute number of people who put a cross against *unimportant* is designated by A_{nw}. The column *unimportant* is always shown in percentages. These figures refer to the respective numbers of people who have actually responded to the single topic (A^*).

$NW = A_{nw} \times 100 / A^*$.

The calculation of the *time of realisation* turns out to be more complicated than that of technical competence. In terms of the estimated realisation period, it was possible to choose between eight columns (until 2000, 2001-2005, 2006-2010, 2011-2015, 2016-2020, 2021-2025, later than 2025, not realisable). The absolute number of those people who put a cross in column 8 ("never"), is abbreviated as A_{nr}. In the data sheets and the diagrams and tables of this volume, column 8 (never realisable, NR) is shown in percentages. These figures refer to the numbers of persons who have actually responded to the category "time of realisation" (A_n), i.e. have marked one of the columns 1 to 8 with a cross:

$NR = A_{nr} \times 100/A^*$

where $^* A_{nr} \leq A^*$

The overall distribution of the individual assessments - not including those persons who absolutely rule out realisation - is described using three statistical parameters. The mean assessment is not defined by the average value, but by the median M. This is produced if you arrange all values according to their size. It then lies in the middle, i.e. by this time, half of the experts expect a realisation to occur. (In the case of data which exists in the form of a frequency table, the frequency of the categories and periods is also taken into account, with the result that an exact date and not just an interval can be calculated.) In order to depict the mean variation of the assessments, in each case the 25 % point Q_1 and the 75 % point Q_2 (i.e. the times by which 1/4 and 3/4 respectively of those taking part expect realisation) are also determined. The results of columns 1 to 7 of the realisation period, in each of which there was a choice of a five-year period, are shown in graphical form in percentiles (fig. 2.5-1).

The left-hand line of the "little house" represents the lower quartile (25 % percentile Q_1), the apex represents the median (M) and the right-hand line the upper quartile (75 % percentile Q_2). The transparent "little house" represents the result of the first round, the grey "little house" that of the second round, i.e. the final result. The basis of the calculation is the number of persons who have responded to columns 1 to 7, i.e. have chosen one of the five-year periods. In the data volumes (internet, diskette), the results of the specialists with the greatest degree of expertise are itemised separately under the "little houses". The line represents the interval between the lower and the upper quartile, while the black triangle defines the position of the median (only second round results). In the tables of this book, the years according to the calculation formula are normally given directly.

Figure 2.5-1:   Graph of the time of realisation

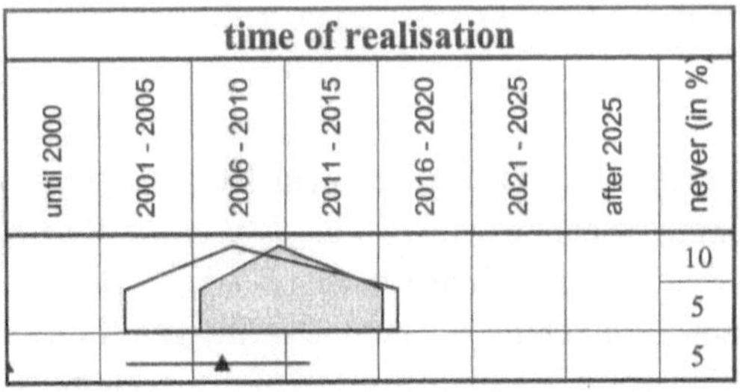

time of realisation							
until 2000	2001 - 2005	2006 - 2010	2011 - 2015	2016 - 2020	2021 - 2025	after 2025	never (in %)
							10
							5
							5

For the quartile and median values, the calculation formulae are as follows:

n (i) = Number of responses for the five-year period i

i = 1: up to 2000
i = 2: 2001 - 2005
i = 3: 2006 - 2010
i = 4: 2011 - 2015
i = 5: 2016 - 2020
i = 6: 2021 - 2025
i = 7: 2026 - ∞

The total of all persons contributing to the distribution is:

$$\sum_{i=1}^{7} n_{(1)} = N$$

The number of persons at the benchmark figures of the distribution is:

$$N_{Q1} = N/4 \qquad\qquad N_M = N/2 \qquad\qquad N_{Q2} = (N \times 3)/4$$

Following the finding of the benchmark figure (M, Q_1, or Q_2) using the formula

$$\sum_{i=1}^{(s-1)} n_{(i)} < N_{M,Q_1} \text{ or } Q_2 \leq \sum_{i=1}^{s} n_{(i)}$$

an interval i and the benchmark figures (decimal adjustment) of the distribution are produced as follows:

$$Q_1 = 1996 + 5 \times (i-1) + 5 \times [\, 1/\, n_{(i)} \times (\, N_{Q1} - \sum_{i=1}^{(s-1)} n_{(i)})]$$

$$M = 1996 + 5 \times (i-1) + 5 \times [\, 1/\, n_{(i)} \times (\, N_M - \sum_{i=1}^{(s-1)} n_{(i)})]$$

$$Q_2 = 1996 + 5 \times (i-1) + 5 \times [\, 1/\, n_{(i)} \times (\, N_{Q2} - \sum_{i=1}^{(s-1)} n_{(i)})]$$

For further information, see the first German Delphi Report or Cuhls/Kuwahara (1994). All remaining categories such as the *highest level of R&D*, the *important measures* and the *follow-up problems* are calculated in a similar way to *importance*. What should be noted, therefore, is that due to the possibility of multiple choices figures of over 100 % can also occur if the total of the entire category (e.g. all important measures) is added up.

The comments were recorded and analysed qualitatively or used for interpretation purposes. No attempt was made to analyse them quantitatively.

As far as *megatrends* (chapter 5) are concerned, the following evaluations can be found: In the first place, people were asked a question regarding their *agreement* with the respective megatrend. This is identified in percentage figures which refer to the number of responses to this specific topic. As only those people who agreed with the assumed megatrend were supposed to continue with their response, the categories of *expected period of realisation* and *influence on science and technology* refer only to those people who agree with the megatrend. The *expected period* is calculated in an equivalent fashion to the previously mentioned *time of realisation* in the technical section of the survey (for the formula see above). The *influence on science and technology* is identified in percentage figures, i.e. the total from the four columns which can be marked with a cross is 100 %.

3   Aggregated Assessments across all Subject Areas

The purpose of this chapter is to provide a summary of the overall results of the German Delphi '98, to bring out the most important general trends and thus also to provide a comparative yardstick for the separate subject areas and topics in the following chapter 4.

3.1   Expertise of the experts taking part

The aim in selecting the experts who were approached was to identify a large number of people with as high a level of expertise as possible in the respective fields. However, the intention was also to include people with so-called "secondary" specialist knowledge within the group of those surveyed, in order to produce a certain corrective counterweight to the scientists with the highest levels of expertise. In the final analysis, an average of 10 % of the respondents classified themselves as experts with the highest level of specialist knowledge, i.e. they work in the area concerned. A further 30 % indicated medium-level (i.e. previous work in the area and/or reading of the primary literature) and 60 % low-level specialist knowledge (reading of the secondary literature and in contact with experts).

At first sight this may appear unsatisfactory but it should be remembered that the 12 questionnaires cover such a wide range of content that the individual expert can testify credibly to a *high* level of specialist knowledge only where very few topics are concerned. Furthermore, it should be noted that the proportion of experts with the highest level of specialist knowledge in the first round was still 13.5 %. Probably several "experts" no longer participated in the second round and/or downgraded their self-assessment. A fundamental tendency to overestimate one's personal specialist knowledge could not be observed on the basis of individual questionnaires and is therefore not reflected in the aggregated result.

3.2   Importance of the Delphi topics

In the course of the refinement of the questionnaires of the first Delphi study, the originally one-dimensional category of importance was split into different dimensions which are indicated in figure 3.2-1. The diagram illustrates the high economic relevance of the Delphi topics surveyed. In fact, this category was crossed in 60 % of cases. However, in social development and in the solving of ecological problems also the experts expect progress in a quarter of the topics. In 20 % of cases an expansion of human knowledge and consequences for work and employment are expected as a result of the realisation of the visions. Whereas economic influences are relevant for all subject areas, albeit with somewhat variable degrees of importance, in the other categories of importance there are some considerable differences which are explored in the respective chapters. Meanwhile, the assessments of the experts from business and science with regard to the different dimensions of importance correspond to a large degree. The assessments of the "types" identified from the megatrends are similarly homogeneous; the location optimists on average view everything as somewhat more important, while those who are sceptical of progress in principle attach a somewhat lesser degree of importance to the topics (for details see chapter 5).

Figure 3.2-1:   Shaping of the dimensions of importance

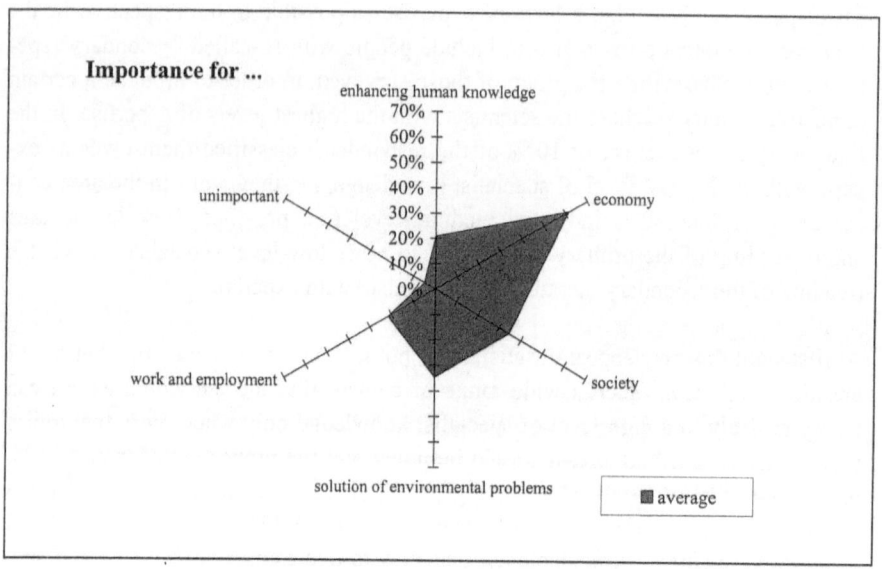

One general point of criticism of the procedure in the first Delphi study was directed against the wholesale takeover of the Japanese catalogue of topics. It was said that some of the questions relevant to Japan were not relevant to basic German and European conditions. Consistent with this, the category "not necessary" was crossed at that time in 10 % of cases and by 10 % of the experts taking part. In the second Delphi study, the total of the figures for importance was more than 150 % (because of multiple mentions the 100 % limit can be exceeded), while less than 5 % of the visions were regarded as unimportant. Both results underline the fact that the selection of the 1,070 visions made by the experts surveyed reflects subjects which are on the whole relevant and thus in this respect the generation of most topics by German experts was successful.

3.3 Expected time of realisation of the visions

According to figure 3.3-1, in its general form the distribution of the realisation periods of the topics cannot be distinguished from the distribution in the first Delphi study. The sole exception is that it has been shifted into the future by around five years. Virtually half of the topics submitted for assessment can, in the opinion of the experts questioned, be fulfilled in the period between 2006 and 2010, a further third between 2011 and 2015. By the year 2000, according to the average assessment of the experts, there would have been *not be a single* realisation. But after 2025, not even 30 visions are categorised. Consequently, the target time horizon of up to 2025 was observed in the majority of the topics.

Whilst no difference in the evaluation of the fulfilment period can be determined between the experts in companies and universities, slight differences are evident between the various "expert types". Thus the environmental pessimists and the population optimists regard the innovations as achievable between 2007 and 2015, while those who are sceptical of progress expect their realisation on average one year later - no great difference against the background of a time horizon of 30 years.

Figure 3.3-1:   Period distributions - Delphi '93 and Delphi '98

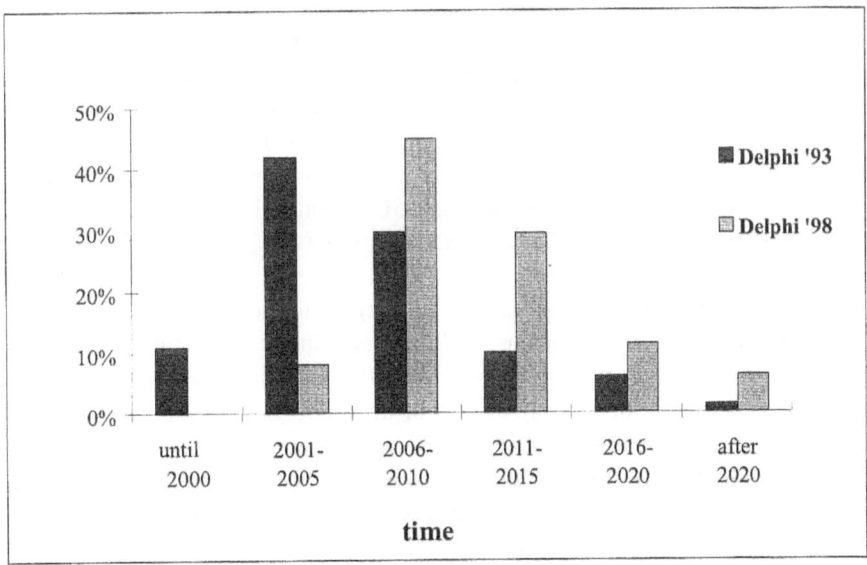

As the average realisation periods are not completely independent of the composition of the topics in terms of the stages of the innovation process, it is also appropriate to discuss them at this point. At almost 40 %, those topics dominate which refer to already marketed, but not yet widely available products and services. A further

30 % cover visions relating to the first economic application of technologies which are already developed and exist in prototype form. The development of prototypes is the subject of a further 18 %, and only somewhat more than 11 % of the topics can be identified with basic research. Thus, in comparison with the first German Delphi study, a more pronounced trend towards application is noticeable.

3.4 Germany's position in research and development: self-perception

In the face of the growing intensity of competition on all world markets in the meantime, it is necessary to identify national strengths and weaknesses, in order to select promising strategies for the future both on a state as well as on a private sector decision-making level. In one category of the Delphi study, a question was asked about the country whose R&D efforts to make possible a realisation of the respective future vision have progressed the furthest. A joint assessment prevails between the experts in the companies and those employed in the universities that the USA is at the forefront of R&D, ahead of Germany and Japan. The representatives of the universities assess Japan to be somewhat stronger. All of the above-mentioned "types" recognise this basic ranking. Only the population optimists see Germany's R&D position as being somewhat more positive.

The cross-comparison over all subject areas in figure 3.4-1 makes clear that Germany is assessed as the leader in the areas of Environment and Nature, Mobility and Transport and Energy and Resources. Compared with the rest of the world, the R&D position in Germany in Space, Services and Consumption as well as Information and Communication, on the other hand, is considered to be insufficiently developed. At least, this is how the German experts see it.

The assessment of the Japanese experts with regard to the identical topics differs from the ranking of the innovations fields only marginally compared with the German assessment. However, Germany was not listed in the 6th Japanese Delphi study as a separate country but merely as part of the entire European Union. A direct comparison cannot therefore be made of how all German items are perceived by the Japanese. Only the 323 topics which are identical in the Japanese and the German studies can be compared on an aggregated basis at the level of the "subject areas". Moreover it remains unclear whether, due to the "distance of Japan", only the economically strong countries of the EU or all of them in their entirety including the weaker countries were used as the comparative yardstick. In the German survey this was clear, as the question was phrased more precisely.

Figure 3.4-1:   Germany's R&D position from the viewpoint of German experts
                 (due to multiple mentions the 100 % mark is exceeded)

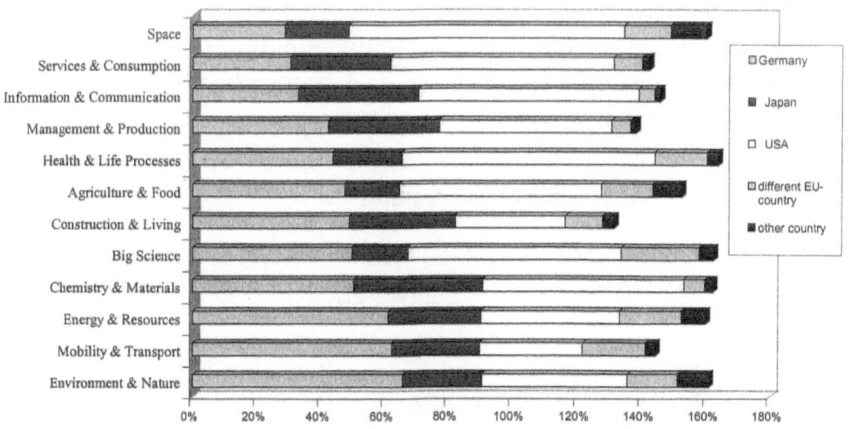

In any event, it is noticeable that the experts in both countries each give themselves better marks. Both groups of experts - the Germans as well as the Japanese - regard the USA as being particularly strong in research and development. After that comes their own R&D in each case. Do the experts therefore have a tendency towards national-centredness? This must be assumed, but cannot be verified from the study due to the varying delimitation of the countries and groups of countries.

3.5   What measures should be taken?

In contrast to the first Delphi study, in which questions were asked about impedi-
ments to the realisation of the future visions from which, indirectly, it was possible
to derive problem-solving strategies, in the second Delphi study various measures
were suggested directly. Although the analysis of individual topics is necessary for
specific recommendations for action, general assessments of the package of measures and recommendations for technology policy arising from them can already be
derived from the result standardised over all topics.

Figure 3.5-1:   Evaluation of the measures

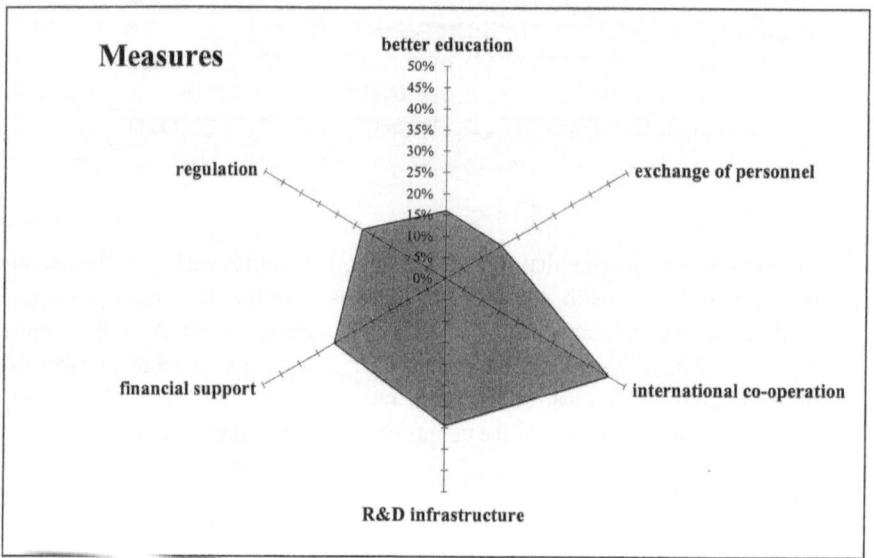

Figure 3.5-1 makes clear that the consequences of globalisation - already apparent
in the first Delphi survey - are also being perceived by the experts surveyed now.
For international co-operation is regarded by them as the most important measure in
order to move science and technology forward. A further emphasis should be put on
improvement in the R&D infrastructure, directly followed by support by a third
party. It is surprising that the often-denounced over-regulation, the inadequate
training situation and the under-developed exchange of personnel between industry
and science are mentioned more as secondary starting points for possible measures.
Compared to the experts from the companies, staff from the universities attach
somewhat greater importance to better training, to a better R&D infrastructure and
increased support by a third party. The location optimists make a similar evaluation
of the measures, while the environmental pessimists demand international co-
operation in a more emphatic manner.

3.6   Which areas of life are threatened by problems resulting from innovations?

One new category compared with the first Delphi study refers to possible follow-up problems which could arise in connection with the realisation of the product visions. In approx. 20 % of the responses, negative consequences are expected for the environment. Conversely, in 20 % of cases, solutions are also expected with regard to ecological problems. This indicates an ambivalent significance of the visions for the future situation of the environment. Negative implications for the various aspects of personal safety rather play a lesser role, while the fear of resultant socio-cultural problems is expressed in 40 % of the responses.

The strong correlation of the assumption of being able to solve the respective problem with the resultant problems can also indicate that the distinction in terms of content between the positively-worded dimensions of importance and the negative follow-up problems was perceived only incompletely by the respondents. In individual cases, however, it must be noted that despite the assumed importance of a solution to a problem such a solution can in turn trigger other problems.

Significant differences between the experts from the companies and the universities exist only with respect to the resultant problems for the environment, which those affiliated with the universities expect to a greater degree. What is surprising is that the *environmental pessimists* and those who are sceptical of progress on the whole expect less resultant problems than the *location optimists* and the group of neutrals (see chapter 5). Consequently, no "extreme" assessments of the matter result from the basic attitudes of the people.

3.7   Identical topics in several subject areas

The remarks made above have already indicated the international consistency of the results. Due to the cross-sectional nature of some visions, they were included in the German Delphi study in two or sometimes even in three questionnaires. This procedure now makes it possible to take a look at whether the different expert communities in Germany have the same opinion when faced with an identical proposition or whether differences are evident in the assessments. For reasons of clarity, only the assessment of the time of realisation was used for comparison purposes. In table 3.7-1, the seven topics with the largest differences in the assessment of the realisation period are listed.

Table 3.7-1:   Dual topics with the highest period differences

Innovation Field	Topics	Median 1	Median 2	Differ-ence
1. Energy 2. Construction	Through advances in rational energy utilisation, long-term energy or heat storage and use of renewable energy sources, en-ergy-independent new buildings and houses are common.	2020	2014	5.9
1. Energy 2. Mobility	By "dematerialising" transport flows (e.g. rationalising road haulage, teleservices for classic supply functions of private households), the development in industry and traffic are sepa-rated.	2015	2009	5.7
1. Chemistry 2. Health	Computer simulations are used to predict the in vivo activity of substances from their structure.	2016	2012	4.3
1. Health 2. Information	Ultramicro-biosensors based on biochemical reactions are ap-plied for medical purposes.	2008	2012	4.1
1. Energy 2. Construction	A heating and cooling system with a heat pump utilising solar energy is used in Germany.	2011	2007	4.0
1. Space 2. Big Science Experiments	A system of optical and other interferometers is set up in space, producing a resolution equal to the apertures of inter-planetary dimensions.	2020	2017	3.9
1. Chemistry 2. Energy	Solid matter electrolytical fuel cells with an output of several 10s of megawatts are in general use for regional power-heat link-age (heat and electricity supply) also in decentral electricity sta-tions.	2017	2020	3.3

First of all, it is necessary to stress that of the 38 period comparisons only seven show a difference of more than three years. If one looks at the accuracy of the estimates, a large degree of agreement prevails between the groups of experts. The exceptions listed, however, underline the fact that the experts in the innovation field Energy and Resources, when compared with the experts in the other subject areas, constantly have a more pessimistic attitude, which incidentally is not reproduced in their attitudes to the megatrends. It can therefore be assumed that this pessimism relates to the specific contents of the topics and themes and not the general world view of the energy experts.

3.8   What lies ahead? An overview

The aphorism: "No wind is convenient for the sailor who does not know the port into which he sails!" is attributed to Seneca (the Elder). The point is, that those whose goals are unclear are unable to utilise the dynamics and driving forces to move forward. The primary goal of the report presented here is to contribute to an understanding of the goals of science and technology. Unlike predictions of the structure of the new government made on the basis of election night exit polls, the future is unpredictable. Neither can current and previous developments simply be extrapolated. Instead, what is involved are the expectations held by specialists – in other words, by those experts in the areas of science and technology whose activities help determine our technological future. An exchange and, perhaps, even negotiations regarding such goals is to be carried out across the borders of individual specialised areas, fields, and topic areas. To do this, however, requires specific materials that can be studied in order to avoid speaking of a "negotiation aggregate". Questions are not only there to be answered, but first of all to be posed, as another anonymous aphorism has it. Asking the "right" questions can already lead to many solutions and answers.

The Delphi report is a representation of the future, neither complete nor comprehensive, and certainly not one that will satisfy all individual interests. Any sorting by specialised area is to be avoided in this view, and the theses as well as the assessments of the surveyed specialists should be allowed to speak for themselves in a kaleidoscopic fashion. This makes selection more accessible than if the report were based on a strict systematic process such as that on the preceding pages. It is impossible to categorize more than 1,000 individual specialised theses in a brief, easily comprehensible manner that is both free of doubt and unassailable. However, despite all these caveats, the attempt to do so is still worthwhile.

Thus, in this view, we will be looking at individual visions of the future, and will describe the extent to which they can provide us with detailed answers to the following questions. Some answers will only appear in later chapters, others will not arise until the course of subsequent discussions.

What is of outstanding or general importance?

What can we expect? The topics, theses or statements in these chapters were selected according to their general importance for both the economy and employment, as well as for society, the environment, and science. These theses were then bundled to form clusters of important visions. The wording of the theses as they appeared in the questionnaire is shown in italics in the subsequent text.

Figure 3.8-1:   The time horizon of the most important innovation fields

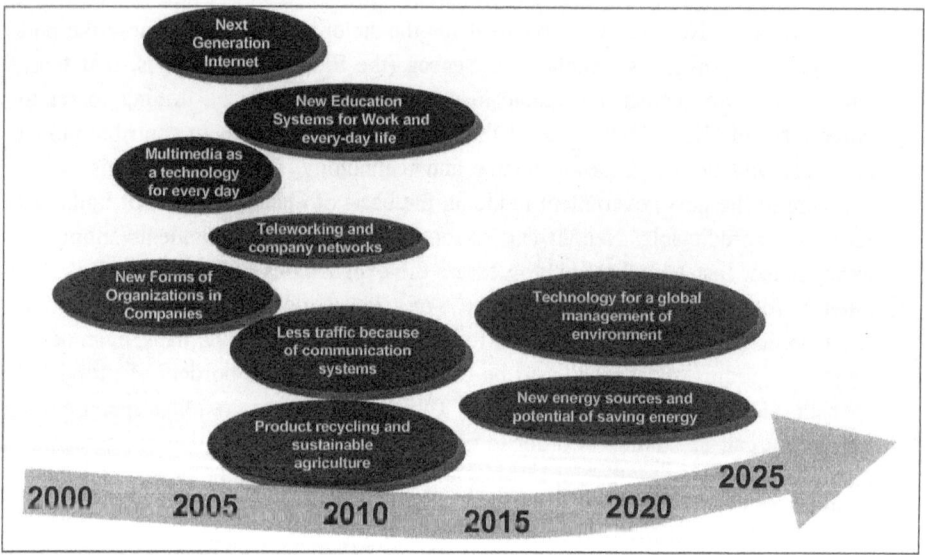

New forms of inter-corporate organisation

In the near future, corporations will cooperate more closely with one another. In the area of research and development, this will also lead to *corporate cooperation that includes input from customers and institutes as a result of the increasing time and cost intensities of R&D projects.*

Based on everything we know, the significance of the employees will increase through *the formation of independent, autonomous areas of responsibility, in order to promote* their *identification with changing corporate goals.* For this reason, *the assumption of responsibility* by employees *of defined portions of the process chain* will become *a scientifically-grounded management goal of personnel development. Identification with individual projects is more important for the purposes of motivating employees than an identification with the corporation, and will thus become a problem to be addressed by top management.* The compensation system will be adjusted to reflect these developments with *that portion of the wage based on work results no longer being based solely on the performance of the individual, but rather on the performance of the group or the overall corporate performance.*

From perspective of technology, microtechnology will increasingly expand throughout corporations. *Components able to integrate sensors, controllers, and actuators have practical applications in microtechnology.* This will alter not only manufacturing operations, but also hospitals and other service providers. The experts estimate that this bundle of cited visions should be realised between 2001 and 2007.

Multimedia as everyday technology

The interlinkage by means of *multimedia networks opens new creativity potentials* for meeting future challenges. *Camera equipment for multimedia applications will soon be part of every PC's standard equipment.* Data networks permit *the secure conclusion of legally binding transactions through technical/organisational solutions for precise individual identification and for data security.* This, in turn, also means that *selection and purchase of many goods will be possible directly from the PC thanks to virtual reality and multimedia, and delivery or pickup can occur independently of normal business hours. Trade via networks* will also *become widespread, thanks to the utilisation of electronic bookkeeping or monetary systems.*

Multimedia networks present new possibilities for not only trade and consumers, but *with the aid of computer simulations,* they promote the *active participation of citizens in public decision-making processes (agencies, local governments, etc.), for example, in the area of building alternatives. The entire construction process will be carried out in a networked (tele) information link, to which all concerned can contribute.* Weather reports will become more-and-more reliable, and daily decisions such as what to wear and whether or not to take an umbrella will receive multimedial support as *satellite data will be integrated into the determination of meteorological base parameters (temperature, humidity) of weather prediction models.* Networked multimedia systems are expected to arrive between 2002 and 2007.

Next generation Internet

A next generation Internet will be realised, whose level of security is high, and which is able to transmit information in real time, thus permitting it to be used to provide telephone service and transmit moving images. For this reason, *the majority of private households will send and receive mail electronically.* The necessary infrastructure for this will be provided by an advanced *wide band cable linkage of all households in densely populated areas* (all of this to occur between 2003 and 2009).

Teleworking and networked companies

The *development of multimedia communications utilising the Internet and intranets will result in general office work at home (with the exception of meetings and negotiations) becoming widespread.* Thus, *with the help of telecommunications, 30 % of all office workers will spend two out of five workdays* working *at home.* Teleworking should arrive between 2005 and 2012 – but not for everyone, and not every day. Further, *research and development projects will be carried forward in a complementary manner at spatially separated locations. Technical information and communications solutions might relationships based on direct communications, permitting individual solutions to discreet parts of an overall problem to be brought together.*

In general, *the dominant form of corporate structures will no longer be characterised by a fixed location and a permanent employee base. Instead, its primary business activity will consist of bringing together numerous partial services provided by individuals or specialised companies – including new university/industry research cooperative efforts in – at various locations, all linked together via networks in a common productivity process.* Will such a development also decrease traffic volumes?

Product recycling and sustainable agriculture

In the intermediate future (between 2006 and 2013) we can expect to "wrestle" with questions concerning ecological economies and limiting global environmental problems. The operational mechanisms of the social market economy will need to be upgraded for this. But important regulatory changes ("re-regulations") are also foreseen. For example, *manufacturers of consumer goods with long service lives will be legally obligated to accept the return of their goods at the end of their service life and to dispose of them. This will result in a recycling system that includes planning, production, collection, and recycling or re-use, with the aid of which, a practically completely closed material cycle can be achieved.*

On the other hand, new technical possibilities will also arise. Thus, for example, satellite-supported *geographic information systems will become operational, permitting the agricultural water budget to be managed on a large scale basis. An overall understanding of the various cycles in the biosphere will lead to the development of agricultural and forest management methods that will permit high yields while simultaneously protecting the biosphere cycles. Those services on the part of farmers for which they receive no monetary compensation (e.g. care of the land) will be assessed and taken into account as a production-dependent performance compensation for half of the area utilised for agricultural purposes.*

From a global perspective, the population explosion represents an immeasurable problem. Humans must eat. They require energy. The specialists participating in the survey believe that *socially and ethically acceptable methods to limit population growth will find widespread application. Transgenetic plants with an improved spectrum of components will be* employed *for the production of feed and foods*, in other words, the attempt will be made to find genetic solutions. *The effects of the destruction of tropical rainforests and the expansion of desert areas on the global climate, together with their interactions will be better understood. Satellite-supported information systems will supply high-resolution spatial satellite and digital base geographic data of 70-80 % of the earth's surface. This information will be freely accessible for all scientific purposes.*

Relief for traffic congestion provided by communications systems

Employment-based mobility requirements will probably *be reduced by 20 % as a result of the increase of home offices. Teleworking and video conferences will also result in a noticeable substitution of business-related traffic. The "dematerialisation" of individual transport streams (e.g. rationalizing and bundling goods shipments, tele-services for traditional supply functions for private households) will permit the link between economic growth and increased traffic to be successfully broken*, in other words, continued economic growth can be achieved without an increase in traffic (2007 to 2014).

New continuing education systems for business and daily life

In the intermediate future, *searches and information flow on the Internet as well as communications with international libraries and databases for both professional and private purposes will already account for 50 % of all communication links. Virtual worldwide universities* and *adult higher education institutes that permit training and ongoing education of the population at home are widespread. Systems to obtain multimedia information on demand are decentrally available worldwide in networks and in the most common world languages.* The same also applies to scientific information. *Multimedia dictionaries from all disciplines will find general application. Here, inputting a letter via keyboard, voice, or by using figures will bring up the requisite information.*

All of this will be regulated by *information databases, able to automatically "learn".* Verbal information transfer will also improve as *an automatic language translation system in pocket-sised format that allows individuals to communicate with others without any prior knowledge of their partner's language will come into common use.* This can be particularly helpful during trips abroad.

For the majority of employees, the utilisation of training and continuing education opportunities with the aid of telecommunications is fully integrated into their working day. Companies will have their own systems of knowledge and experience management to which employees will have access via easily understood structures, and in which they can combine various knowledge and experiential elements to form new perceptions. Refresher and training systems for professional development planning of middle-aged and older individuals will become generally available. Here, such individuals will be able to acquire new specialised knowledge and technical qualifications. For this reason, *education will increasingly lead to individual qualification bundles rather than to* generally applicable *degrees whose contents are fixed by educational research and hierarchical assignment.* This group of concepts can be expected to be realised between 2006 and 2014.

New energy sources and potential areas of energy saving

The development of new energy technologies requires time. The required structural alterations to accommodate the distribution of these new energy sources is a slow process. The Delphi experts all agree that there will be increased use made of solar energy, but put the necessary advances at the end of the second decade of the next century (2013 to 2023). Only then will *the proportion of renewable energy used to generate electricity (hydroelectric power not included) in Germany* exceed *10 % (today, approx. 0.5 %).* Also at that time, *multilayer or laminate solar cells capable of achieving energy conversion performance levels of more than 50 % will become practical.* In addition, *a process by which water is disassociated by solar radiation will become available. Solar-thermally driven Stirling motors will find widespread application to generate electricity in sun-rich countries. System costs for network-coupled photovoltaic systems will drop below 4,000 DM/kWh (currently: 15,000 DM/kWh).*

But *fuel cells based on solid polymers with power/heat couplings* will *also be widespread in residential buildings. Industrial and commercial operations* on the other hand, will *utilise decentralised, high-temperature, fuel cell equipment to simultaneously generate electricity and high-temperature heating.*

New technologies will help save energy. But we must still wait nearly two decades for this to occur. *Many new procedures for replacing today's energy intensive processes in primary industries will* then *become practical, so that the specific energy requirement for basic industrial processes will have sunk by one third compared with today's levels.* Some of the areas specifically cited include, *for example, steel and nonferrous metal casting processes that are near final tolerance, membrane technologies to replace thermal separation processes, structurally rigid anodes, microwave and pulse drying. Up to 50 % of the traffic in the European Union will be replaced by high-performance, high-speed vehicles – both for passenger and freight transport. Traffic-related environmental stress in German cities will be re-duced by 30 % thanks to structural changes in municipalities.* Among these changes will be the lifting of the current separation between living, working, and shopping spaces.

Technologies for global environmental management

Ecosystem research on closed systems (biosphere) will improve our understanding of the global ecosystem to the point where a global framework of basic conditions for human survival can be created. Techniques to landscape deserts will be applied throughout the world to stop the desertification. Drought and salt resistant strains of agricultural plants will be developed with the aid of biotechnology. These plants will be able to produce high yields, even in areas where the water table is threat-ened by salinisation. Cultivation in biomasses will reduce the water requirement for agricultural plants. Agricultural areas – including ones that will be newly devel-oped in former deserts and above the arctic circle – will be planned and managed by locally specialised land utilisation offices. (This entire scenario may take until 2014 or even 2024 to develop).

Which controversial Utopias will we not see?

Our kaleidoscope only goes as far as 2025. Some future projects will not be realised by then. In some cases, they may never come into being. Thus, as many specialists note, we should not expect *3,000 m high residential or commercial buildings in which up to 50,000 individuals live and work, to be built in super-megacities with populations of 10 to 20 million.* We should be spared this, even though 600 m high buildings for more than 5,000 inhabitants have already been realised today.

Neither will staying in one's home for recreational activities become common. Quite the contrary, our democracies will see no radical limitations placed on indi-vidualism. *An annual mobility quota for motorised travelers for trips to or around the city,* remains an Utopian idea, although an *eco-bonus in the form of a tax rebate for ride sharing* is foreseeable. Neither *will the voluntary limitation to automobile usage* be rewarded *by special bonus procedures such as private electricity rebates.*

Politicians and economists can all breath easier! The communal system, that was only abandoned in the East a decade ago, will not even be introduced in the West. The idea that, for example, *in cases of verifiable incompetence the management of a corporation will be removed and replaced with a workers' committee,* will remain unfulfilled. An *"electronic parliament" (electronic state senate) that meets and op-erates in televised sessions so that decisions regarding proposed laws (regulations) can be made with the aid of electronic voting by the constituents,* will probably remain an utopian idea. Even *the extensive privatisation of governmental functions* will reach an end. *Public institutions to ensure safety, such as the police, corrections departments – with the exception of the judicial branch – will* not *be privatised.* Neither – based on the results of the survey – need we expect to see *the introduction of a federal medical insurance structure based on a catalogue of treatments*; this will not occur.

Innovations for economic development

The initial glimpse at the possible future has already indicated several interesting trends. This second glance focuses on innovations that are not necessarily on the top of the list, but that become more apparent when questions touch on specific areas. Let us start with concepts that will have a significant, overall influence on the economic development of Germany and the world.

New organisational structures between corporations
New quality standards in food production
Satellite-supported traffic control
Electronic currency as the payment method in multimedia networks
Photonics and new chip generations
Satellite technologies
New materials and processes
Bio- and food technologies

New organisational structures between corporations

New models of cooperation that *permit the corporation to concentrate on its core business, thus allowing manufacturing and development activities to be largely shifted to the provider industry* are vital for economic development. Thus, *the ability to efficiently carry out complex projects – including research and development – in cooperation with numerous companies* will become a new *core skill.* So as not to loose the overview of all these parallel activities, *all relevant corporate data and the most important information about the corporate management environment will be condensed, and be presented and updated on a daily basis to support management of the corporation.* According to the experts, these concepts concerning the corporation of the future will be realised between 2001 and 2008.

New quality standards in food production

New quality standards will arise in the area of food production. *Core data concerning farm animals will be gathered* (2002 to 2006) *on an individual basis (cattle, pigs) and on a species basis (poultry, aquaculture), to be systematically evaluated (feeding, growth, illnesses, prophylactic procedures, treatments). In order to determine the genetic source of plants and animals with the goal of protecting investments in biotechnology, molecular probes and genetic fingerprinting techniques will find practical applications* (2002 to 2008).

Satellite-supported traffic control

Between 2003 and 2008, *satellite navigation systems with increased and expanded performance, and under international, civil control will become routine aids for efficient air, sea, and land traffic.* In particular, *a satellite-supported, online information system for those traffic areas particularly sensitive to weather (i.e., the shipping and air industries) employing onboard data processing by the user will become practical.*

Electronic currency as the payment method in multimedia networks

As *trading in networks* that have now become *more secure as a result of data encryption techniques continues to expand through the employment of electronic accounting or monetary systems* (2002 to 2005), *digital currency* (2002 to 2007) will become increasingly more important with respect to means of actual payment.

In the intermediate future, new technical ideas will become economically more important. These will be found into everyday products where, however, they may not always be visible.

Photonics and new chip generations

Between 2006 and 2013, the infrastructure of multimedia systems will be switched to optical systems. Thus, *optical solution transmissions will be employed for fiber optic transmissions over long distances, e.g., for intercontinental cables on the ocean floor. Fiber optic materials permitting transmission speeds of 100 GB/s will be developed (currently: 20~30 GB/s).* Photonics will also become increasingly important for data storage. *Optical memories with storage densities of more than 100 GB/cm² will appear* (2007 to 2013). Finally, *new display technologies such as "luminescent plastics" (that can be shaped into any desired form) will be ready for practical applications* (2007 to 2013).

The current, dramatic developments in chip technology will continue. *By utilising the development potential above the chip level (chip-on-board technology, multi-chip modules, 3-D chips, and wafer scale integration), jumps in packing density of two orders of magnitude will be able to be achieved (2006 to 2010). Both nonvolatile and editable RAM of more than 100 GB will become practical (2007 to 2014). A technique to permit large-surface, linkage semiconductor, single-crystal layers to be produced on glass plates will also be developed* (2010 to 2018). *Finally, LSI circuits with switching times under 1 picosecond will become available* (2006 to 2013).

Satellite technology

Aside from those applications of satellite technology already mentioned (e.g., weather forecasting), there will be progress in the technology of satellites themselves. Thus, *the supply power for satellites' digital electronics will be reduced from 5 V to less than 1.5 V, permitting their power requirements to be reduced by one order of magnitude* (2004 to 2009). *Wherever thermally possible, structural elements will, as far as possible, be eliminated, thus reducing satellite mass by one order of magnitude* (2005 to 2014). In addition, *operating in a hybrid mode, components made up of a combination of semiconductors and high-temperature superconductors will appear* (2007 to 2014). Since *satellite communications systems for wide band transmissions at more than 1 GB/s per circuit will be developed* (2005 to 2011), quicker applications will be possible. However, improvements in current application systems are also expected, e.g., a *unified, interference-free, global satellite navigation system with an accuracy of half a meter* (2004 to 2009). These developments will permit, for example, a *practical, all-weather system for aircraft takeoffs, landings, and ground movement* (2005 to 2010).

New materials and processes

The chemical industry will continue to make advances in the development of new materials. *Polymer fibers with a tensile strength of 40 % of the theoretical value and an elasticity moment of 90 % of the theoretical value (2006 to 2013)* will be developed. *"Smart" materials with the ability to adapt to external influences will be employed in greater numbers to increase the efficiency of machines (2005 to 2011).*

Between 2004 and 2010, the following new processes will appear, to make the manufacturing of new and improved materials more cost effective. *A new process will be developed that allows tailor-made polycondensation materials to be produced. Mass production of semi-finished goods made of compound materials such as carbon fibers will be advanced, making such materials widely available as inexpensive, light-weight construction materials. Further, a technique to sinter solid state, nanoscale particles at temperatures in the 800°C range will be developed, leading to heat resistant, high-performance materials (e.g., ceramics) based on SiC or Si_3N_4.*

New catalytic systems will also be developed to permit epoxies to be manufactured by direct oxidation (2006 to 2014) and *make selective CH-activation in methane a reality, so that methanol can be produced directly from methane* (2008 to 2015). In addition, *a new refining technique will be developed that will allow titanium to be produced as cheaply as aluminum* (2010 to 2018). *In technical syntheses, the reaction and material separation steps will be process-integrated in a single device, for example, by means of reactive distillation or in a membrane reactor* (2006 to 2016).

Effective and efficient testing techniques will become increasingly important. These will permit, for example, *the condition or performance reserves of metallic materials to be examined in non-destructive test procedures, from the results of which the probable, residual operating life can be determined* (2005 to 2013). *Non-destructive tests to detect cracks of less than 10 μm in ceramics will also find practical application* (2003 to 2009). Further, *there will also be advances in software certification techniques, which will permit nearly error-free software to be rapidly developed on a large scale* (2007 to 2015).

These new processes will permit *production technology solutions that allow manufacturing according to customer specifications to be carried out at nearly the same cost as series production* (2003 to 2010). This will also make it possible, *to generally halve the current time required between the generation of a new product and its introduction into the market* (2004 to 2010). All of this will have significant effects on economic development.

Bio- and food technologies

Cell cultures in large-scale bioreactors will find widespread employment for the manufacture of highly pure substances die (e.g., pharmaceuticals, high purity chemicals, proteins) (2005 to 2011). *Transgenetic plants with an improved spectrum of components will be employed for the production of feed and foods* (2006 to 2011). *By utilising the most up-to-date food technology, small to medium-scale producers will be able to successfully bring a wide variety of innovative products to the marketplace* (2004 to 2010). Despite widespread concerns regarding the risks of genetic technology, these concepts are reassuring for a market economy, particularly in view of the fact that, until now, there was never any assurance that innovative, medium-sised operations could participate in the advances in the area of bio-technology.

Which economically significant innovations will probably never be realised?

In many cases, innovations that could lead to a drastic and economically significant savings in energy, research and development will reach their limits – at least based on current knowledge: In other words, they will probably not be realised within the foreseeable future. However, opinions tend to diverge on this point. We feel that, if 20 % of the Delphi experts say "never", while the remainder indicate a time period beyond 2025, hope for the realisation of the innovation in question should be abandoned! This, in turn, is an assessment of our knowledge of the future.

It will probably not be possible, *to manufacture automobiles equipped with 2 litre ceramic engines that produce low pollutant levels without the aid of catalytic converters*. This does not, however, mean that 2 litre automobiles are not generally possible. Instead, it appears that the ceramic engine without a catalytic converter is the sticking point in this instance. *A new refining technique that permits titanium to be produced as cheaply as aluminum* is also felt by many experts to be impossible. The same applies to the development of *a technique to produce iron from ore that uses only 70 % of the current primary energy requirement*. Neither is an economic jump in superconductor technology resulting from materials expected that can operate without special cooling: The experts doubt that *a superconducting substance whose transition temperature lies within the normal temperature range will be developed*.

Limits are also seen in the area of information and communications technology that will continue make humans with their intellectual capacities indispensable, just like *portfolio managers are for portfolio management*. Neither will *real time reporting be likely to replace the periodic examination by regulatory authorities*. Because of inadequate international coordination and technical options, neither will it be possible *to charge a copyright fee on an international level for every graphic or image scanned in or retrieved from multimedia systems*. Neither does the structure of Germany's political system permit, *all supports or subsidies for the agricultural industry to be done away with so that only large-scale agricultural operations, based on free-market terms would still operate in Germany*.

While there is no question that "digital" currency will become increasingly significant, the hopes of those who feel that the trees of the information society grow into cyberspace will be dashed. This is because the visions that will probably not be realised includes the idea that, *because of electronic home shopping, up to 50 % of all food items will reach the consumer without having passed through the hands of traditional merchants*. Digital currency simply does not extend that far. The Euro and the cent will continue to circulate, long after the Mark and the penny have disappeared.

Innovations for work and employment

Those innovations that will become important for work and employment in the near future all affect the previously outlined corporation of the future, and will therefore not be repeated in detail at this point. Instead, we will focus on expected developments in the intermediate future.

More flexible working hours
New corporate structures
Work-intensive services

By 2015, there are expected to be a number of interesting innovations that will have extremely significant effects on work and employment. The most important of these is that *public employment services will gradually be replaced by private employment agencies* (2003 to 2009). The experts believe this will improve the efficiency of job placement, but not that there will necessarily be more jobs available for the agencies to work with.

More flexible working hours

Additional innovations are characterised by an increased flexibility in working hours. *In this way, operating hours can be largely divorced from the employees' working hours because the fault rates of automation solutions* will be very low (2005 to 2012). For this reason, *the operating hours of highly capital intensive production operations can, in most cases, be expanded to an average of 20 hours per day* (2003 to 2009). *Modern forms of organisation* will permit *more than 40 % of the employees of efficient corporations* to be made up *workers with term contracts or job contracts* (2005 to 2013). Despite attempts to make working hours more flexible in Germany, *standardised administrative and office routines will be performed by competent personnel in developing nations with the aid of tele-working* (2002 to 2009). On the other hand, there exists a threat of a dearth of highly qualified workers, thus forcing *productive companies to win qualified personnel by offering family-friendly employment times* (2003 to 2010).

New corporate structures

More flexible solutions will also be found in the area of compensation, with the *basis for the payment of all employees being split 50/50 between payment for time worked and payment for work results* (2007 to 2014). In addition, *objective calculation formulas will be applied where it will no longer be quantitative factors that determine level of that portion of compensation based on work results, but instead, qualitative factors (schedule maintenance, errors, etc.)* (2003 to 2010). Finally, *the majority of companies will compensate workers with stock or shares, because this has proven to be a method that increases motivation and therefore productivity* (2003 to 2010).

In addition, *traditional, functional departmental divisions will be replaced by structures oriented towards product lines or specific customers* (2003 to 2009) which, in turn, will lead to an increase in available jobs. *The simultaneous existence of differing forms of organisation and operating principles at a single production site will also promote the production flexibility and customer-closeness demanded by the market* (2003 to 2009) and will thus promote growth. *Small to medium-sized companies in particular will bundle their employment and sales activities based on models of inter-company cooperation in order to have greater weight in the marketplace* (2003 to 2010). Finally, *the criteria for technical and economic effectiveness when making decisions regarding the use of automation or human labor for manufacturing will be determined with the aid of highly efficient aids* (2004 to 2012). The result: More security in the workplace.

Work-intensive services

New forms of services will arise whereby *a significant portion of the working population will take advantages of services ranging from dwelling to "hotel" services* (2008 to 2016). Further, *retail businesses will, to a greater degree, come to assume the form of wholesalers as new information and economic, as well as organisational possibilities become available.* To a certain extent, this may see the return of the *"errand boy"* (2003 to 2009). Thus, the future of employment to some extent lies in models from the past – far-ranging questions based on new ideas. In the intermediate future, employment possibilities continue to be closely linked with the organisation of the economy, not, it should be noted, with any decrease in wage costs. This is because, again in the intermediate future, *there will occur a decentralisation of decision-making responsibility to the operative level for all those decisions for which the information basis is greatest at that level* (2003 to 2010). This creates work!

Which concepts will not appear in the working day?

Since the innovations are expected to provide few impulses for work and employment in the long term, we shall provide a brief overview of the most important innovations with their implications for work and employment, which, based on the current views of the experts, will probably not be realised. Unfortunately, *a production system that takes into account and makes allowances for the limitations of older and (mentally and physically) handicapped persons* will probably not become widely established. Neither *will the dual professional education in industry become so attractive that individual faculties will be forced to compete for students.* Despite a number of *well-developed communications strategies in new organisational forms, hierarchically motivated instructions will not become superfluous.* Equally Utopian are *creativity breaks for all employees in leading-edge companies representing 5 % for all employees in general to 20 % of the working time for R&D personnel.* It may be a small comfort for construction workers in an industry currently suffering greatly from unemployment that, in the future, *buildings will* probably *not be constructed by fully-automated methods that do not require human intervention.*

Innovations for social development

After looking at the most important theses affecting the economy, work, and employment, let us now turn to those innovations that will have a particular effect on social development.

Disarmament monitoring with the aid of modern satellite systems
Emergency management with new information systems
Nutritional information: ambivalence
Child- and senior-citizen-friendly buildings and dwellings
Innovations for a more handicapped-accessible environment
Advances in medicine

Disarmament monitoring with the aid of modern satellite systems

In the short term, the experts expect only a limited influence by technological innovations on society. Nonetheless, even as East/West tensions diminish, there still exist immense weapons arsenals. *Satellite and other space technology will be employed* in the near future *to monitor compliance with disarmament agreements* (1999 to 2005). According to the experts, some of these are already being utilised today.

Emergency management with new information systems

Health emergencies in the civil sphere are also at the threshold of decisive breakthroughs necessary in order to service the civilian population. Medical services will soon become more effective and efficient through the introduction *of functional emergency management (ambulatory) to provide the fastest possible acute treatment for stroke victims in centres for ambulatory operations* (2002 to 2007). In the intermediate future, *image and information transmission systems between emergency vehicles and hospitals will become increasingly widespread for emergency situations* (2003 to 2009).

Nutritional information: ambivalence

The experts disagree as to whether genetically altered food products need to be identified. The Delphi experts see a social advance in the fact that *improved tracking methods for generic procedures or products manufactured by methods of genetic technology will provide adequate methods of intervention (listing policies, boycotts), that will force trade organisations to require food manufacturers and processors not to employ any suspect products or methods* (2002 to 2007). In the long term, *despite improved and less expensive tracking methods, the requirement to provide food information will not reach the point where a detailed listing of every ingredient and process used in the manufacture of a given food or beverage must be supplied.*

In the future, the experts feel that still only a minority *of consumers will possess an adequate knowledge of health and nutrition to be able to understand the informa-tion about food, and to react accordingly.* Neither do they feel that *the widespread industrial processing of foods and beverages will greatly limit the current variety in the organoleptic quality of nutrition, nor that there will be a "trend towards uniform consumption and uniform taste".* However, *the addition of synthetically produced substances to food due to improved production techniques for natural materials will probably* not be *prohibited.*

Child- and senior-citizen-friendly buildings and dwellings

In the intermediate future, the *child friendliness of housing developments and buildings* can be determined taken into account during the planning and construc-tion stages of new projects by means of *more objective assessment procedures and standards* (2006 to 2014). *The integration of aspects of daily life such as housing, work, recreation, and shopping in a single complex of buildings* will ease *communi-cations between different generations, offers new opportunities for identification with one's own environment* (2008 to 2019). *Organised housing exchanges will* also *include housing reserves of, e.g., single, elderly individuals, without forcing them out of the immediate housing area* (2005 to 2013). In the short term, *variable con-struction methods will permit individual dwellings to be uniquely designed, even in multi-complex structures* (2002 to 2007).

Innovations for a more handicapped-accessible environment

The increase in the numbers of the elderly in our society will lead to an increased requirement for care. For reasons of cost alone, this requirement cannot simply be met by increasing the number of caregivers. To ease communication, *a technology to transform media, e.g., for a system to translate sign language, will become prac-tical, so that hearing impaired individuals can communicate with the non-hearing impaired over great distances* (2009 to 2016). *A portable speech generator that allows handicapped individuals to convert their desires directly into the spoken word will become generally available,* (2008 to 2016). Even *systems to guide and orient the visually impaired on sidewalks and walkways by the use of sensors will be employed in Germany* (2009 to 2019).

Self-sufficiency will become easier *as dwellings equipped with robots and other devices will be generally available. Here, older individuals and the handicapped will be able to support themselves without the need for special caregivers* (2009 to 2021). When such self-sufficiency is no longer possible, the *technology for the home care of the elderly and ill can be shifted to different special and environ-mental conditions* (2005 to 2011). *Technological adaptations and the design of the home environment will even develop into a separate service branch* (2005 to 2011) with potentially positive effects on the number of available jobs.

Alternatively, *micro-association communities* may also *arise, and assume the re-sponsibilities for the care of individuals in cases of illness or degenerating social circumstances. Such communities will operate as associations with their work based on accepted scientific standards of care* (2008 to 2015). In the intermediate future, even *spatial distribution and the design of apartment buildings will be able to be adapted to meet the changing needs resulting from the various phases of life of their inhabitants or to accommodate generational changes without any great effort* (2007 to 2016). In support of these changes, *intelligent wheelchairs able to automatically adjust to stairs, escalators, and rising/falling paths will become practical* (2007 to 2013).

Advances in medicine

Despite the great advances of the past, medical science has yet to free humans from all ills. In the intermediate future, some breakthroughs are, however, expected. *Various micro-devices able to move independently within the body will become standard in clinical practice (e.g., for blood diagnoses and the treatment of throm-boses)* (2010 to 2018). Again in the intermediate future, *the pathogenesis of Alz-heimer's disease will be understood* (2006 to 2013) and *an effective treatment will be developed* (2011 to 2019). *An AIDS treatment that will permit the progress of the disease to be halted in its early stages and with which the long-term effects can be overcome will become clinically available* (2005 to 2011). *Effective vaccines against HIV, the causative agent of AIDS, will be available in the affected develop-ing nations* (2010 to 2019).

In the intermediate future, *methods based on genetic analysis to predict individual risks of developing genetically related illnesses such as cancer and high blood pres-sure will become widely available* (2006 to 2013). It will even be possible *to iden-tify those genetic groups at risk from diabetes, high blood pressure, and arterioscle-rosis (typical adult illnesses that are inherited), thus allowing the molecular causes of these diseases to be determined* (2008 to 2014). *Parkinson's Disease and other diseases of the basal nervous system will be understood to the point where treat-ment will be able to alleviate all symptoms associated with these illnesses (trem-bling, imbalance, orientation disorders)* (2011 to 2018). Further, *an effective insulin preparation that is administered orally, will be developed* (2005 to 2011). Even *the neurochemical mechanisms of alcoholism and its genetic components will be under-stood* (2010 to 2019).

Even if *education of the population, training of medical personnel, and measures to increase confidence all contribute to increase the number of organ donors to the point where the need for donor organs can be completely met* (2006 to 2014), recovery will still not be possible in many instances. In these cases, *recognised procedures to predict the course of the illness in question will probably be available, thus allowing the extremely elderly or totally disabled to make informed decisions regarding euthanasia* (2008 to 2017). In addition, *the "morning after" pill will have been adequately investigated and become ethically acceptable on a widespread scale* (2003 to 2009): Progress for society.

Innovations to solve ecological problems

Despite high levels of unemployment and empty public coffers, solving ecological problems continues to represent one of the major tasks of the future. Those theses with the greatest significance for the solution of ecological problems are presented below.

Innovations to protect the atmosphere
Innovations to protect the oceans and groundwater
Innovations to protect forests and land areas
Resource protection through recycling

Very few proposed solutions offer any hope in the short-term. Exceptions are *speed limits for automobiles*, that have been discussed for years in Germany. Based on environmental and consumption model calculations, these *have been determined to be 120 km/h on, 80 km/h on secondary roads, and 30 km/h in urban areas* (2001 to 2008).

However, a greater number of solutions to various environmental problems are expected to be found in the intermediate future. In particular, with the aid of information technology, the sources, extent, and intensity of pollution will be able to be determined and reduced. *A global network for monitoring environmental stress throughout the world – particularly the spread of environmental problems – will become generally available. This system will receive data in real time, 24 hours a day, and will integrate this information, systematically analyse it, and then distribute it throughout the world* (2008 to 2015).

Innovations to protect the atmosphere

Since motor vehicle emissions are a significant source of atmospheric pollution, a reduction in harmful emission products can be expected in the intermediate future if *a technology such as diesel catalytic converters, particle traps, lean NOx catalytic converters, or very precise combustion techniques are generally applied to new types of vehicles* (2007 to 2013), *thus bringing the concentration of nitrous oxides in exhaust gases significantly below the EU-mandated levels for 2005* (2006 to 2011). *Vehicle technology will also permit a reduction in the precursor substances that result in the formation of photosmog to an extent that there will be a reduction in peak ozone values of 70 % compared with 1990* (2009 to 2017). *The introduction of new materials that make vehicles lighter and stronger will reduce fuel consumption still further* (2004 to 2010). Finally, *zero-emissions vehicles will be developed as local areas in which only such vehicles may operate are mandated* (2004 to 2011).

But the areas *of building technology and insulation will also make such progress that tomorrow's dwellings will require only 20 % of the energy currently used for buildings* (2005 to 2011). Among other things, this *will reduce world-wide carbon dioxide emissions by 20 % of the 1990 value* (2010 to 2020). This would be more than the amount fixed in the 1997 Kyoto conference on measures to prevent global warming. Further, *alternates for CFCs and halogens will be employed throughout the world (including developing nations), so that no further enrichment of these substances in the atmosphere will take place* (2006 to 2013). Finally, *new agricultural techniques will permit rice plantings with drastic reductions in the amount of released methane* (2007 to 2014).

Innovations to protect the oceans and groundwater

Protection of the world's oceans can be improved by *monitoring methods (e.g., long range monitoring systems), that are so effective that any dumping in the ocean can be immediately registered and the culprits can be identified* (2003 to 2010). *A world-wide monitoring system providing operational, near real time, total surface monitoring of oceanic pollution (in coastal areas and on the high seas) and of ocean currents will become reality* (2011 to 2017). *Biotechnological or chemical/physical processes will be available to permit oil spills to be effectively removed after tanker accidents in a manner that protects the environment* (2006 to 2012).

It will be easier to safeguard the groundwater supply since the *separation of so-called "utility" and drinking water will be widespread in homes* (2007 to 2015). In addition, *compact biotechnological water treatment systems that can be set up at decentralised locations will also be developed, and will be capable of removing not only nutrients but also substances that are more difficult to break down from waste water* (2005 to 2011). Groundwater pollution resulting from heavy metals or organic pollutants *will be handled by the use of biotechnological and physical processes to permit on-site treatment of hydrocarbons, halogenated hydrocarbons, other organic compounds, and heavy metals* (2004 to 2013).

Innovations to protect forests and land areas

The earth's forests will be better protected by *the removal of large-scale threats throughout the world (in particular, to tropical rainforests and conifer forests at high latitudes in Canada and Russia). Tree felling will instead be replaced by selective lumber management processes* (2009 to 2018).

Based on new syntheses, the biological decomposition of herbicides, pesticides, and fungicides will be significantly improved, thus contributing to the protection of land areas, particularly those used for agricultural purposes (2004 to 2010).

Resource protection through recycling

In urban areas, organic and combustible waste will be almost completely included in a recycling circuit (composting, plastic recycling, paper recycling, gasification for material recovery) (2007 to 2014). New types of information systems can also make effective contributions to the protection and savings of resources. Thus, *information systems that accompany products will find widespread use, detailing how, at the end of its service life, a product can be broken down and its component parts can be gathered and resources recovered from them* (2007 to 2014). *As an example, only constructions will be used in the automobile industry that permit automatic disassembly and complete recycling* (2006 to 2012).

According to the opinions of the experts, concepts concerning an *increase in the average utilisation of all vehicles by 50 % through the active arrangement of ride-sharing groups* remain ecological Utopian ideas. Neither does the current view see any *agricultural crop growing methods that do not require the use of herbicides for raising cereal and oil crops*. Efficient energy consumption will reach its limits in areas such as the development of *lights that convert up to 75 % of electricity into light energy*. In comparison, a conventional light bulb converts a mere 5 % of current into light. There are also grave reservations concerning the *"safe and secure" enclosure of Plutonium $_{239}$*. Finally, the experts do not foresee great success in the realisation of a technology that *converts the CO_2 in exhaust gases from power plants and other large sources into fuel such as methanol by the addition of hydrogen*. Not every problem lends itself to a technical solution. Instead, other ideas are required that, for example, prevent the creation of hazardous substances from the outset.

Innovations to expand human knowledge

Since the Delphi study has taken on the future of science and technology as its subject, the question must be asked, which basic areas, aside from economics and society, can expect to see elementary breakthroughs that lead to an expansion of human knowledge and understanding. For this reason, the following topics that are of great significance with regard to an expansion of human knowledge are presented here.

Space exploration **Exploration of the microcosm**

Space exploration

Since basic research tends to be a long-term theme, the specialists foresee only limited initiatives for the expansion of human knowledge in the near future. In the intermediate future, thoughts are directed at space. The surveyed experts expect *the question of dark matter in the universe to be solved* (2007 to 2020). *The electromagnetic radiation coming from space will also be able to be measured across all frequency ranges. In particular, the energetic and locational accuracy in the x-ray range will be greatly improved, and the current gap in the gamma ray range between 1 to 100 GV will be closed* (2006 to 2014). It should also be possible to *carry out investigations of Mercury with the aid of satellite probes* (2011 to 2019).

Significant gains in knowledge from space research are also expected in the long term *as a result of probes to the outer planets beyond Saturn* (2011 to 2021). *Optical and other (x-ray) interferometers will be set up on the moon or in space to study the coronas of stars and the aurora emissions of extrasolar planets. This equipment will provide angular resolution accuracy of up to 1 micron of an arc second* (2022 to post-2025). *Returning probe missions to many bodies in the solar system, and employing miniaturised vehicles–* as have been indicated by the most recent Mars mission (although no physical samples were returned) – will be carried out (2013 to 2023).

A system to detect gravitational waves will be developed and will employ space technologies and structures or constructions equipped with highly accurate acceleration sensors (such as frictionless or resistance-free satellites) (2012 to 2020). *The equivalence principle and the relationship between space/time curvature and matter will be experimentally investigated* and researched (2011 to 2023). *The cause of the violation of the symmetry principles, responsible for the imbalance between matter and antimatter in the universe, will be determined* (2009 until post-2025). In other words, in 20 years we will know much more about space than we do today.

Exploration of the microcosm

Aside from looking at far distant worlds, research into the building blocks of matter itself appears promising. *A general theory concerning the smallest particles and their interactive forces, that goes well beyond the current standard model, will be developed and experimentally confirmed* (2008 to 2019). *Leptons and quarks, cur-*

rently thought to be the elementary building blocks, will prove to be made up of even more fundamental particles (2004 to 2014). Within the same time period, *the cause for a proton's rotational pulse will be discovered* with the aid of electron and proton beams *(beam/target experiments)*. *A new source for ultracold neutrons with high intensity will be developed to study the fundamental properties of neutrons (life span, electrical dipolar moment, etc.) and the processes of quantum mechanics* (2011 to 2020). *Neutrino detectors capable of identifying individual neutrinos will be designed* (2001 to 2011). *Quark/gluon plasma will be discovered and investigated as a new form of matter* (2007 to 2018). *Electron/positron colliders with energy levels above 500 GV will become operational to permit unanswered questions in the field of particle physics to be investigated in* (2010 to 2019).

Particle research will also be advanced as a result of *the problem of the creation of matter from elementary particles and hadrons being solved* (2011 to 2021).

In the area of biology, the mechanisms governing the formation of neuronal networks will be defined at the molecular level, as will the *structure and function of all molecules participating in the signal transmission process* (2017 to after 2025).

Technical simulation of many vital processes is questionable

Humans will attempt to technically recreate a number of vital processes. Many of these advances would result in large increases in human knowledge. However, not everything is possible. Thus *technical systems able to reproduce themselves based on the model of living organisms* as well as *processes to preserve living organisms on the basis of the hibernation principle,* will presumably never eventuate.

Neither can human intelligence be reproduced, as a result of *the mechanisms of creative performance in humans being so well understood that they can be utilised in the area of information technology.* Neither *will computers implanted in the human brain with the aid of electrical or electromagnetic processes be able to tap into the information stored there. The development of hybrid intelligence that joins ICs and living cells, will probably also not be possible.*

4   The Subject Areas

The following chapters are divided up in line with the subject areas. The structure of each subject area is laid out in a comparable manner. In view of the wealth of material, however, not every detail can be shown in each case. The selection of the key topics may appear arbitrary, but is based on particularly clear-cut relationships or subjects which have received special comments from the Delphi experts. This should invite you to identify your own key topics and to set them within your own personal frame of reference.

As indicated in chapter 2.3, the individual topics comprise a verb which describes the so-called stage of the innovation process of the subject. The stages of the innovation process range from basic research (something *is clarified*) via development (*is developed*) and the first application (is *in practical* use) through to widespread dissemination (*in general use/widespread use*). This should be taken into account.

The more participants assess a topic, the more sound is the result. With regard to the evaluations, only the results of those "experts" are taken into account who work in the respective field (high level of specialist knowledge) or know a lot about it (medium or low level of specialist knowledge), i.e. are familiar with the primary or secondary literature.

4.1   Information and Communication

(contributed by Dirk M. Harmsen)

Structure of the innovation field

The innovation field of Information and Communication is particularly broadly-based. The range of 111 topics extends from the basic technologies of information and communication techniques via aspects of processing and presenting certain information, security aspects of technology use, applications of communication techniques between people, and between man and machine through to questions concerning the future "knowledge society".

If one looks at the stages of the innovation process to which the 111 topics can be assigned, around 23 % of them are to be found in the area of clarification and development, approximately 44 % of the topics are devoted to the first commercial applications and 33 % deal with the widespread use of the innovations.

Around 61 % of the topics in this field of inquiry agree with those of the 6th Japanese Delphi survey carried out over the same period - the highest agreement ratio of all twelve subject areas.

Who was surveyed?

In the first round of this Delphi survey, 287 people filled in the questionnaire in this field of inquiry; in the second round, the figure was 206. Of these, 34 % are employed in universities and 27 % in companies, 12 % can be assigned to private non-profitmaking institutions and 7 % to the civil service. 17 % are employed elsewhere and 2 % gave no details.

The vast majority of the respondents have a low opinion of their own specialist knowledge - presumably due to the considerable breadth of the subject range covered in the questionnaire. This does not mean, however, that the individuals could not assess the topics in a sufficiently informed manner. Only in the case of two topics is the proportion of the experts with a high level of specialist knowledge roughly 20 % and thus comparatively high:

- *As a result of the development of multimedia communication using the Internet and Intranet (e-mail, WWW, telephone conferencing etc.) general office work at home, except for meetings and negotiations, is very common.*

- *The majority of all private households send and receive electronic mail.*

Both topics were assessed by a very large number of experts (190 and 196), meaning that the results are very "stable", because the more responses there are, the more meaningful is the statistical parent population.

The two topics with the lowest response rate (number of respondents: 45 and 47) are: *communications systems for secret and extremely complex information are developed on the basis of the principle of optical chaos synchronism*; and *micromachines are developed for the medical field which use the ATP in the blood as energy sources.* The proportion of the respondents with a low level of specialist knowledge here is 82.2 %, and 85.1 %, and the proportion of those with a high level of specialist knowledge 0 % and 2.1 %. These topics are so specialised that probably hardly anyone among the Delphi experts works in the field or that the experts could not be identified via the usual database classifications because they work in a "niche" of a research field. (To identify the experts, see the methods volume.) However, a check would have to be made as to whether any specific research is being carried out on it in Germany. This applies to all topics which, compared with the other subjects in the field, show significantly lower response figures.

Why is Information and Communication an important field?

With regard to the question, for which reasons the innovations mentioned in the topics are needed, economic development is mentioned most frequently, considering the average of all responses, followed by social development (fig. 4.1-1). The creation or the preservation of jobs occupy the third rank in terms of the importance of the innovations in the field of Information and Communication. In the view of those surveyed, these obviously make only a very scant contribution to solving the ecological problems.

Figure 4.1-1:   What are the innovations in the field of Information and Communication important for?

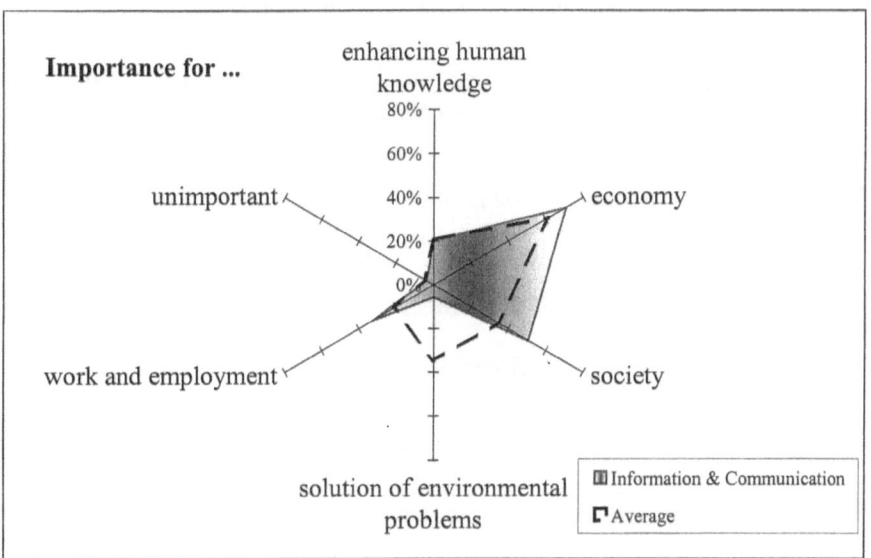

When will the topics become reality?

In the view of the experts, around 11 % of the innovations will occur between the years 2001-2005, approximately 46 % in the period 2006-2010 and around 32 % in the period 2011-2015. The innovation which has been to be possible at the earliest time is the *replacement of conventional cameras by digital cameras:* the average figure estimated by the experts here is 2004, with a range of variation of ±3 years (upper and lower quartile).

According to the assessment of the experts, the three innovations with the lowest likelihood of realisation are that after clarification of *the physiological and psychological mechanisms that cause human beings to make mistakes, a behavioural alarm system to warn people before they make mistakes* is applied, that *with the aid of electrical and electromagnetic procedures, computers can record the information stored in the human brain* and that *the pattern recognition performance of man is being reached (even caricature).* In these cases 41.1 %, 37.0 % and 36.4 % of the respondents think that the innovations are never achievable.

What is the level of Germany's research and development in Information and Communication?

The Delphi experts were asked which country leads the world in research and development, the USA, Japan, Germany, another country within the European Union or another country. In the opinion of the experts surveyed, the United States of America today occupy the first place in the field Information and Communication, followed by Japan and Germany (cf. fig. 4.1-2). Other countries within or outside the European Union only play a leading role in a few areas. Incidentally, the Japanese experts also assess the country ranking in a similar manner.

Figure 4.1-2:   Germany's R&D position compared internationally

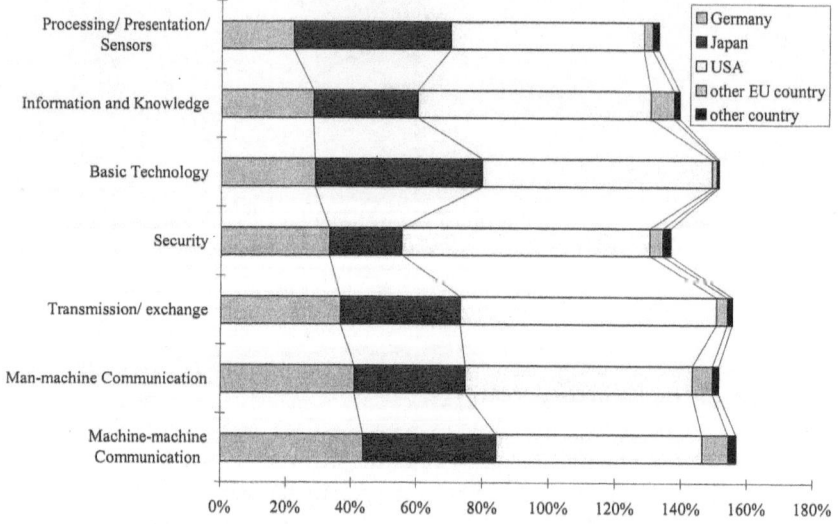

In the view of the German experts, the Japanese hold a clear lead in research and development in the field of display techniques, e.g. in

- *high-resolution, large-format 90-inch colour flat screens which at the same time form a wall decoration,*

- *touch-sensitive flat screens in the form of a desk surface,*

- *a communication system, in which an entire wall is used as a monitor, in order to create a closer link between office branches and thus, for example, to convey the feeling of "presence",*

- *flexible, robust (folding, roll-up) displays,*

- *handwriting recognition technology (e.g. one with which over 99 % of hand-written Chinese characters can be read).*

In addition, leadership is conceded to Japan in

- *fully-automatic production in the mass consumption sector (up to the level of the car),*

- *the use of intelligent robots which have sight, hearing and other sensory functions, can assess the situation in the outside world for themselves and make decisions independently, and*

- *three-dimensional television which can be viewed without glasses.*

The experts regard Germany as having a clear leadership in research and development only when it comes to micro-machine technology and electronic fee charging for road traffic: *Components which integrate sensors, controllers and actuators come into practical use in micro-machine technology*, and *Charging systems for use by trunk roads (e.g. including GPS) are in general use*. In the latter area, a high R&D status is also accorded to other EU countries.

What needs to be done for Information and Communication of the future?

The question was asked as to what measures should be taken to make the innovations become reality, e.g. better training of staff, increased exchange of staff between industry and science, international co-operation, a different R&D infrastructure, support by a third party, changes in regulations or other things. In the subject area of Information and Communication, the experts named in first place international co-operation, followed by the improvement of the research infrastructure and support by a third party (fig. 4.1-3). Better training and more rapid technology transfer between science and industry are in fourth place. The necessity of a change in regulations (more regulations or less ordinances and laws) is considered to be slight.

Figure 4.1-3:   What measures should be taken?

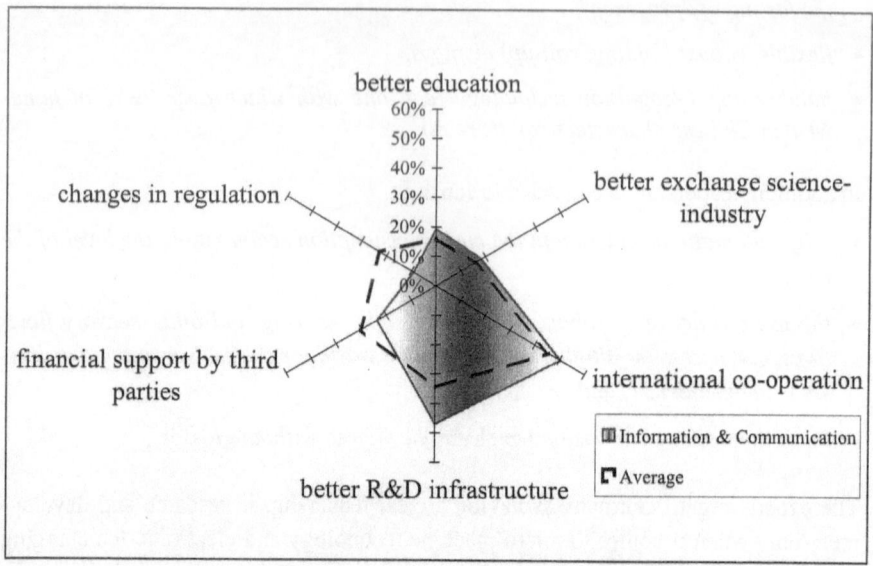

With regard to possible resultant problems which may occur in achieving the innovations, in first place the experts name social and socio-cultural problems (56 %), in second place problems of security such as the safeguarding of privacy or the misuse of personal information (19 %). On the other hand, there is very little expectation of any resulting problems for the environment.

Key point: In Germany a rationalised information and knowledge society will be established

In the subject area of Information and Communication, the experts foresee positive effects on the creation or on the preservation of jobs due to the realisation of the envisaged innovations only to a restricted degree. Even in the near future, up to around the year 2010, further opportunities for rationalisation are opening up thanks to information and communication technology, to the extent for example that *production in the mass consumption sector (up to the level of the car) will take place on practically a fully-automatic basis* or *remote maintenance systems for plants and machines* will make surveillance staff nearly unnecessary.

Whereas in production the main focus is on the potential for further job reductions, information and communication technology is responsible for more of a qualitative change where administrative and service activities are concerned. Because *due to the development of multimedia communication using the Internet and Intranet (e-mail, WWW, telephone conferencing etc.) general office work at home, except for meetings and negotiations,* is possible. The technical basis for this is universal *broadband cabling* and *a next-generation Internet offering high security which can transmit information in real time, with the result that a telephone service and the transmission of moving pictures are possible.* The virtual proximity of staff who are physically separated and *the feeling of "presence"* is created by *a communications system in which an entire wall is used as a monitor.* A loss of jobs must be expected in particular where translation and secretarial staff are concerned, if *language and voice recognition* are *in widespread use as general means of input to the man-machine interface between terminals and users* together with *automatic translation equipment.*

On the other hand, does information and communication technology also safeguard old and new jobs? There are opportunities for this in the education system, which is undergoing far-reaching changes due to information and communication technology. As, in the future, *for the majority of employees the use of educational services with the aid of telecommunications (telematically-supported further and continuing education)* is to be *integrated completely into the course of occupational working time,* the qualification level of employees is being continually adapted to new requirements. In principle, presumably, in not even 10 years from now *thanks to remote teaching systems training and continuing education of the population will be possible from home.* In addition, *education can increasingly lead to individual qualification packages instead of to degrees,* so that entry to a profession should become easier. These new possibilities reduce training times and costs, and thereby lead to a greater supply of young, highly-qualified workers who are well-equipped for the labour market of the future. However, these opportunities for the education system also have to be seized by using the new means of information and communication and integrating them into an efficient education system.

In essence, the Information and Communication sector will increasingly manifest itself as a global field in which, without international co-ordination and in the absence of a decision for or against a certain research structure, no inroads can be made into the American-Japanese domain. On the other hand, the influence of information and communications technology on general economic development will continue to increase, as a result of which working life and society as a whole too will be subject to massive changes.

4.2 Service and Consumption

(contributed by Dirk M. Harmsen)

Structure of the innovation field

The range of the subject area, covered by 78 topics, extends from electronic financial services, electronic payment transactions and teleshopping via mobile information and communications aids and transport services, care and help services, the "intelligent house", leisure activities and educational aids, new services such as electronic voting through to disaster prevention and safety issues.

Of the 78 topics just under 3 % belong to the innovation stage clarification and development, approximately 28 % of the topics are devoted to the first commercial applications and 69 % deal with the widespread use of the innovations. 27 % of the topics in this field of inquiry are the same as those used in the 6th Japanese Delphi survey carried out over the same period.

Who was surveyed?

In the first round of this Delphi survey, 215 persons filled in the questionnaire covering this field of inquiry. In the second round, the figure was 163 persons. 48 % of those taking part are employed in companies, 16 % in the civil service, 12 % in universities and 4 % in the private non-profitmaking institutions, 15 % elsewhere and 4 % gave no details.

In this subject area too, most respondents themselves only admit to a low level of specialist knowledge with regard to the contents of the various topics. Only in the case of the proposition that *financial service companies specialise in a narrow spectrum of activity in their core business sector by allocating most administration functions to sub-contracting companies with the appropriate facilities,* is the proportion of the experts with a high level of specialist knowledge more than 13 %.

The two topics with the highest response rate (number of respondents 154 and 153) are *electronic supermarkets in which shopping can be done at any time of the day or night (from ordering to delivery at agreed times)* and: *most "computerisable" services are in worldwide competition. Technology makes possible communication with the customers and the transfer of the service, while face-to-face communication is no longer considered to be absolutely necessary.* In this instance, the proportion of the respondents with a high level of specialist knowledge is 6.5 % and 9.2 %, and the proportion of the respondents with a low level of specialist knowledge 66.2 % and 57.5 %.

As it was difficult to locate "experts" in the field of Service and Consumption, a large part of the persons who were addressed were selected from the service sectors (including banks, insurance companies, transport companies etc.). Because of this, the specialist knowledge of the participants is only high for very few topics (remember "high" means actively working in the field). However, as "consumers" or suppliers of the respective technology, these people believe they can also offer an assessment on other questions, but therefore put a cross against "low specialist knowledge" in greater numbers. This is why these relatively high shares of medium or low-level specialist knowledge result. There is virtually no difference in the assessment of the topics compared with the other fields.

Why is Service and Consumption an important field?

In terms of the average of all responses, the topics of the subject area Services and Consumption are regarded as the most important for economic and social development. Influence on work and employment occupies third plac in the order of importance (fig. 4.2-1). The substitution *of public by private emj 'oyment agencies* is named as the most significant innovation. In the opinion of tho e surveyed, the innovations mentioned in the topics clearly contribute little to th solution of environmental problems or to the expansion of human knowledge.

Figure 4.2-1:   What are the innovations of the field Service and Consumption important for?

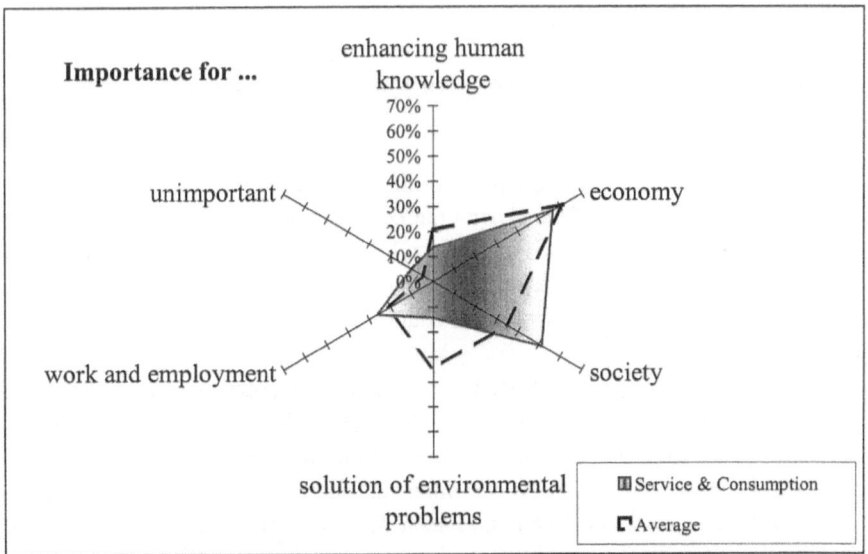

When will the topics become reality?

In the opinion of the experts, around 13 % of the innovations will be realised in the period from 2001 to 2005, approximately 54 % in the period from 2006 to 2010, around 27 % in the period from 2011 to 2015 and 6 % in the period from 2016 to 2020. At the earliest, *the problem of secure financial transactions in electronic banking (also on the Internet) can be solved with data encryption techniques for practical purposes (i.e. with comparable standards of present-day systems).* In this instance, the average figure estimated by the experts is the year 2003, with a range of variation of ±3 years. And it can already be noticed that data encryption technologies improved meanwhile – although of course – problems still have to be solved and the creativity of "hackers" should not be underestimated. The comments of the experts agree mainly at this point.

In the view of the experts, the three most unlikely realisations are:
- *After organisational research has proved their efficiency, the public institutions set up to guarantee internal security, such as the police or the penal system, with the exception of the administration of justice, are privatised.*
- *It has become usual not to have to leave one's home in order to organise one's leisure time.*
- *An "electronic parliament" (electronic state assembly) operates and sits in parliamentary television programmes, so that decisions on draft bills (regulations) are made with the aid of electronic plebiscites.*

In these cases 60.6 %, 59.1 % and 47.3 % of the participants believe that these visions can never be achieved.

What is the level of Germany's research and development in the field of Service and Consumption?

If one looks at the current R&D position in the subject area of Service and Consumption, in the opinion of the experts the United States of America occupies first place, followed by Japan and Germany who both lie equal (cf. fig. 4.2-2). Research and development in other countries hardly play a leading role. The Japanese experts also come to a similar assessment of the country ranking.

In the view of the experts questioned, research and development in Japan leads the world in the utilisation of robots, e.g. in the areas of leisure, care and emergency services as well as in the "intelligent house" and in disaster prevention (*robot leasing offices for various service robots*, robots as sparring partners *for sporting activities*, robots as *guides of the blind*, as *home helps* or those *which recognise people in the event of a disaster, make enquiries and provide help*). World leadership is also accorded to the Japanese in the area of the development of forecasting techniques for natural disasters and in the control of electronic equipment by visual and voice contact: *due to progress in forecasting technology for landslides as well as mud and rock slips, the number of deaths will be considerably reduced worldwide. Thanks to the use of modern sensors, the control of machines via visual or voice contact is part of the normal equipment of simple machinery in industry and the household.*

Figure 4.2-2:   Germany's R&D position compared internationally

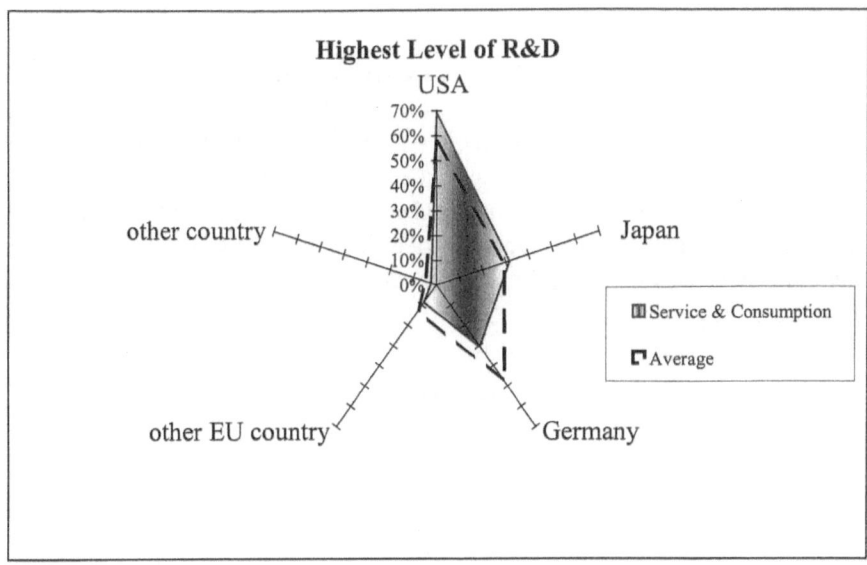

The experts regard Germany as leading in research and development in the field of educational aids (*institutions are widespread, in which re-evaluation (evaluation) programs are systematically organised for traditional trade and for art and culture, in order to contribute to life-long learning. A refresher and training system for the occupational development planning of middle-aged and older people is universally established, in the framework of which they can acquire new specialist knowledge and technical qualifications*), of energy management in the "intelligent house" (*energy management of the new facilities of private houses (sensors, actuators, e.g. shutters, heating/cooling management etc.) by means of a central computer is standard*), care and emergency services (*technology for the domestic care of old people and the sick can be geared to varying space and environmental conditions. The adaptation of technology and the organisation of the domestic environment develop into a service in their own right*) and in smart card applications for transport services (*All public transport in Germany can be used with a standardised contactless smart card, so that tickets are no longer needed, and charging and inspection are henceforth carried out only via the card. Car-sharing systems are equipped with smart cards for billing and GPS systems for location tracking purposes and as a result are widely accepted. Bicycles for hire using a smart card are provided at*

every train station and every public transport bus stop. Charges are demanded for using most public roads in accordance with the individual usage intensity (kilometres, current traffic volume).

In the latter area, a not inconsiderable R&D status is also accorded to other EU countries. The experts look at important research and development activities in other countries which do not belong to the European Union only in relation to the above-mentioned topic devoted to the *development of forecasting techniques for natural disasters.*

What needs to be done for the "service society" of tomorrow?

As the most important measures to achieve the innovations highlighted as a central theme in the 78 topics, the experts (almost equally) mentioned the improvement of the research infrastructure and international co-operation, followed by the necessity of a change in regulations and support by a third party (cf. fig. 4.2-3). Better training and better technology transfer between science and industry are ranked four and five respectively.

Figure 4.2-3:   What measures should be taken?

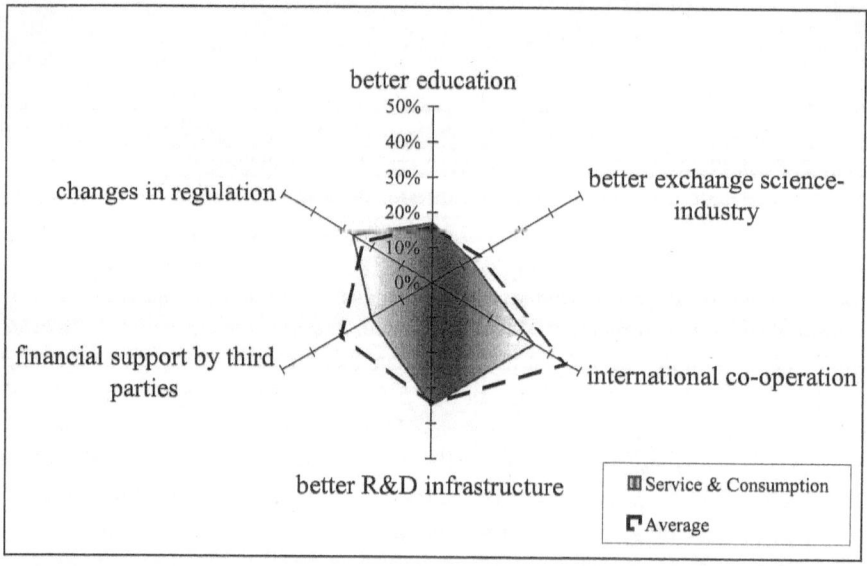

Although it is not a question here of straightforward research issues, it is the state institutions or companies and their R&D facilities which should be more strongly reminded of their duty, not least with regard to highly application-oriented products. The frequent mention of international co-operation clearly shows also that services -

in particular electronic cash, financial services and teleshopping - will no longer develop solely on a national level.

What follow-up problems might ensue?

Possible resultant problems such as environmental problems and security also were brought up for discussion in this questionnaire along with social and cultural-social consequences. In the field of Services and Consumption, the experts expect problems to arise in the social and socio-cultural area (62 %). In no other subject area are the resultant social and socio-cultural problems considered to be so important as in "Services and Consumption". In second place come security problems such as the safeguarding of privacy or the misuse of personal items of information (26 %). Scarcely any fears are expressed of problems arising for the environment (7 %).

Key point: New services will challenge work and leisure in Germany

The social problems arising from IT-supported services have various dimensions. On the one hand it seems that, certainly where the organisation of leisure time is concerned, inter-personal contact will suffer, if *facilities are in common use whereby, using virtual reality, the experiences of travel, film performances, sporting contests and amusement parks can be enjoyed in the homes of ordinary German families* and *it has even become usual not to have to leave your home to organise your leisure time.* However, practically two thirds of those surveyed regard the realisation of this vision to be impossible.

Face-to-face-communication with customers in most computerisable services which are in world-wide competition will decrease, because *technology makes possible communication and the transfer of the service.* In exactly the same way, direct customer contact will be lost, if from the period between 2006 and 2014 *more than 30 % of everyday goods in terms of clothing, food and accommodation are purchased in Germany via teleshopping* (but on the other hand, some experts even doubt that the 30 % mark will be reached at all). Extreme social consequences are feared in the event that *freely programmable machines are installed in houses and hospitals to assist in nursing.*

To the psychological and interpersonal dimension of the ensuing problems are added components linked with political distribution if, on the one hand, *the real world (e.g. travel, sport, shopping)* remains *reserved for a small minority* and the majority are merely offered *a virtual reality such as electronic supermarkets.* On the other hand too, the discriminating effects of *Pay TV* can further reinforce social class differences.

In a broader sense, the security problems of the innovations listed also have consequences for society. In this context, *the problem of secure financial transactions* and *of digital money in electronic banking* despite *data encryption techniques* continues to be perceived as a security risk, while at the same time a high degree of importance is granted to these innovations as far as economic development is concerned. The situation is similar with *the smart card containing information on one's personal financial situation.* There is a fear of a further threat to privacy if *financial transactions* are *permanently monitored to prevent money laundering, insurance policyholders* are *put under surveillance by their insurance companies using telecommunications-supported image transmissions* and the *earth's surface is watched by* private *service businesses with a resolution of 1 to 2 metres.*

4.3   Management and Production

(contributed by Carsten Dreher)

Structure of the innovation field

This subject area contains topics on new materials, aspects of saving resources, on production and operating facilities technology, on quality and safety at work, on control of the supply chain and on aspects of gainful human employment and management. As is also the case with a number of other subject areas, it principally contains topics with first applications and widespread dissemination. Less numerous are topics on the expansion of the knowledge base and on the clarification of technical principles.

Who was surveyed?

More than 45 % of the total of 179 persons taking part in the second round come from companies, a further 37 % from the universities. People from other institutions such as the civil service are more rarely represented. Over 15 % of the experts from the private sector and over 24 % of the experts from the universities themselves claim to possess a high level of specialist knowledge. The representatives of the civil service - in terms of numbers one of the smallest groups - assess their own specialist knowledge the highest. This testifies either to the successful choice of experts by the project team or to varying assessment behaviour within the different groups.

Why is Management and Production an important field?

The importance of the topics discussed shows differing emphasis in comparison with most other subject areas: the significance for economic development as well as for work and employment is assessed as considerably higher. The significance for social development - so the comments of the experts lead one to suppose - is assumed to be as high as in other subject areas due to the employment effects. What is surprising is the general impression of the experts that the contribution of the innovations to solving ecological problems is less than in other fields. Only a detailed comparative analysis can clarify to what extent the new trends in Management and Production support or hinder a clean environment. Per se, at least no contributions are expected by the experts towards solving environmental problems (cf. fig. 4.3-1).

Figure 4.3-1:   What are the innovations in the field of Management and Production important for?

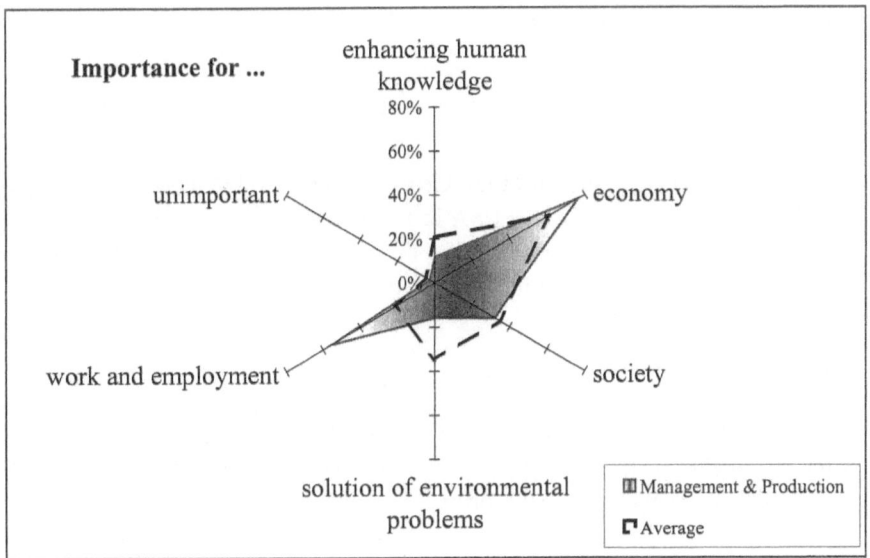

When will the topics become reality?

The experts surveyed assume that of the 71 topics as many as 59 will be realised by the year 2010. The median lies in the year 2008 and in comparison with the other subject areas is thus very early. After 2010 the realisation of just twelve topics is expected. This clearly shows that in the field of Management and Production topics with a short- to medium-term outlook predominate.

What is the level of Germany's research and development in the field of Management and Production?

When taking an overall view of the assessment of the highest R&D position, a pattern comparable to other fields emerges: the R&D position in the USA is judged by the experts to be the highest, followed by Germany again ahead of Japan. However, when a more detailed analysis of the themes is made, significant differences become apparent:

- *In the areas of management and forms of organisation, the outstanding assessment of US American R&D achievements is all too obvious. Here, Japan and Germany are regarded as being equally strong.*

- *In terms of control of the supply chain, i.e. organisation beyond company boundaries, Japan can demonstrate the highest R&D position. The experiences of Japanese industry in collaborating within the so-called "keiretsu", the pyramid-shaped structure of networked financing, information, R&D and procurement associations, must have led to this assessment.*

- *In the opinion of the experts, in areas such as materials and production engineering Germany lies on average equal with the USA.*

- *When it comes to innovations involving environmental protection and quality, the R&D position in Germany is assessed by the experts as being number one in the world.*

If one compares the assessment of the German experts in this field with the Japanese experts, in both countries the USA is assumed to be in the lead. However, the Japanese experts claim "second place" for themselves. The fact that both groups of experts take a "second best" attitude applies in general to the entire Delphi survey.

What needs to be done?

On the question as to which measures are particularly important for the achievement of the innovations, within this subject area a distinctly different profile of measures emerges than in the other fields (cf. fig. 4.3-2).

The focus is clearly on training. Although not specified in more detail in the questionnaire, the comments lead one to assume that this refers not only to researchers but in particular to management and employees. In addition, the interchange between theory and practice is rated as a measure of above-average importance. Changes in regulations are regarded as important, above all where issues of resource-saving and safety at work are concerned and on the whole lead to this measure being mentioned with average frequency. In the opinion of the experts, measures for promoting research or for improving the R&D infrastructure are considerably less important. The demand for state measures is thus distinctly underrepresented, which is hardly astonishing: after all it is the companies which have to test and carry through the application of new management, organisational or production techniques themselves.

Figure 4.3-2:   What measures should be taken?

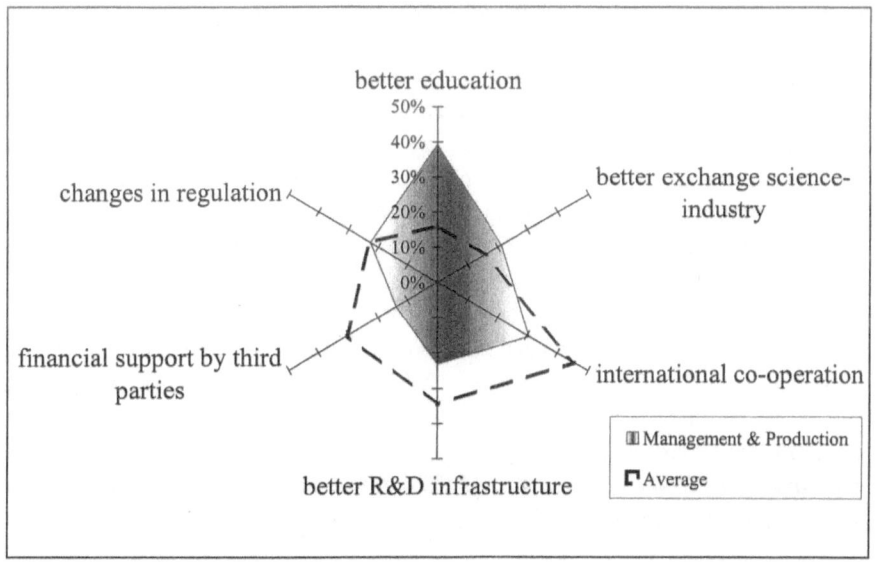

What follow-up problems might ensue?

In terms of possible consequences of the innovation, the experts anticipate social problems above all. If one takes the comments as indications it is, for example, the new forms of organisation which can lead to growing unemployment, to "mercenary employees" etc. Clearly a (further) worsening of the employment situation is expected in the event of the realisation of the topics brought up for discussion in the questionnaire, as the assessment of the megatrends also confirms (74 % of the approx. 2,300 respondents assume that *technical progress and the global redistribution of jobs will increase the average rate of unemployment in the industrialised countries* - as early as the immediate future).

Key point: In the German sphere, manufacturing co-operations and spontaneous networks will come into being

A remarkable central theme runs through the 71 topics and the varied questions from technology, the world of work and management:

In the German sphere, manufacturing co-operations with a high degree of responsiveness will come into being. From a market perspective these are characterised by the fact *that they have specialised in customer groups and no longer on products and try to satisfy them by spontaneous network formation* (realised by 2008; the proportion of the experts who think this is **not** possible is 14.2 %). Your *customers order a product on-line to their own requirements* (2005; 0.6 %). *Apart from as-*

sembly activities, the networks have *taken everything out of stock and transferred it elsewhere* (2007; 17.9 %) and *manufacture at the same costs to customers' orders as they did before in standard manufacturing* (2007; 3.4 %).

Technology has adapted itself accordingly: questions with a technical or chemical background, for instance about the *influence of "intelligent materials" on the efficiency of machines* (2008;0 %), *general use of precision casting or forging* (2007; 2.4 %), *shortening of car assembly by 20 % using the bonding process* (2007; 5.4 %), *multiple mini-reactors for synthesis plants* (2013; 0 %) or *rapid-assembly systems instead of nuts and screws* (2007; 5.9 %) show a tendency to support flexible manufacturing co-operations or employees. As well as this, *production processes for the continuous manufacture of a wide range of products in small quantities* are being used (2007; 0 %) and maintenance is being performed irrespective of location (*remote maintenance systems:* 2008; 4.8 %). *In any event, information systems no longer form a barrier to the choice of organisation* (2008; 5.4 %). In this way, *operating times of 20 hours per day are achieved* (2006; 2.5 %) which, however, *due to the lower susceptibility of the automation solutions to disruption, can be dissociated from the hours of work of the employees* (2008; 6.3 %).

In organisation, everything seems possible: *Thus several principles can exist at the same time* (2006; 0.7 %) or *old forms are replaced in blanket fashion by customer-oriented structures* (2006; 6.5 %). However they appear - it is important that *decision-making is so decentralised that decisions are taken where the information base is best* - i.e. at the operational level (2006; 6.9 %). *Up to 40 % of employees will perform large parts of their work using new technologies remotely or from home without any loss of productivity* (2010; 9.5 %). *Half of their pay is geared to results* (2010; 13.5 %), *the results of their efforts are determined by qualitative factors* (2006; 7.4 %) and *to this end are geared to group or plant performance* (2005; 4.8 %). *The majority of companies will therefore pay with share certificates* (2007; 7.5 %).

Independence will be the keynote and *integral areas of responsibility are expected which encourage the motivation of the employees* (2004; 0.6 %). Nevertheless: the maintenance of the *identification of employees with the company in view of the independent handling of projects* will become a problem of top management (2005; 8.2 %), especially as these "centrifugal" developments can also lead to a situation in which in efficient companies *more than 40 % of employees are given employment contracts or contracts for work of limited duration* (2009; 11.3 %). *Standardised administrative tasks are performed via teleworking by competent personnel from developing countries* (2006; 13.3 %).

The result of the last topics shows that at least as far as the locational independence of the networks is concerned, quite considerable doubts persist (13.3 % of the experts think the topic is not feasible). Even more experts evince scepticism when it comes to the concept of a *Virtual Company* not tied down to location (2012; 14.1 %). There is also a sceptical assessment of other visions: *Family-friendly working hours in order* to gain *qualified personnel* (2006; 10.3 %) or more *women in positions of leadership due to their better network management capabilities* (2009; 12.2 %).

Taken as a whole, however, it is the unity of the topics and the responses to them which impresses. Most fulfilment times (medians) are located in the period from 2006 to 2008. By 2010 at the latest, 75 % of the experts think every innovation will be implemented. The "vision" of future companies consists of many independent and autonomous employees or commissioned self-employed people in highly flexible production systems. Management will principally play the role of information and contact broker. Nevertheless, almost a third of the experts cannot imagine that *new forms of organisation will make hierarchically-motivated work instructions unnecessary* (2008, 28.0 %), or that in the year 2008 things will depend so much on the employee that with his colleagues he *will vote the management out of office in the event of demonstrable and serious errors* (2015, 56.9 %). In the view of the experts things won't actually get that far.

4.4   Chemistry and Materials

(contributed by Gerhard Jaeckel)

Structure of the innovation field

The majority of the topics in this field of inquiry include the areas of materials and new techniques in production and analysis. The area of materials predominates by far. To the extent that the fields of application of the materials are mentioned, they refer to such diverse areas as microtechnology, current storage and current conduction technology, optoelectronics, sensor technology, bonding, dressing and implants in medicine, structural components in transport engineering, engine and turbine construction and heating technology in the building sector. The fields of application of the new techniques in production concern mainly energy provision, environmental protection, the transport sector, agriculture and the production sector in general. The topics on analysis constitute a comparatively small group.

The area of new materials and chemical products which is described here is, in terms of its starting-point, a cross-sectional subject in a line of development which is in principle continuous. So it is only natural that, with regard to the innovation progress of the contents of the innovations, a really balanced distribution results of the stages of the innovation process considered in the survey - a distribution which in part diverges considerably from the other subject areas. Thus, the topics on research and development and those on the applications in products and processes are represented in approximately equal numbers.

The range of the subject area is exceptionally large and includes practically the whole of research and development and many fields of application which are also covered by other subject areas of the Delphi survey. "Highlights" must therefore be located first and foremost in the individual topics, if one simply disregards the "hot" application subjects of energy, transport and the environment.

As far as straightforward materials topics are concerned, those on functional materials predominate slightly over those on structural materials and thus underline a trend which is also reflected in the contributions made at the large international conferences over recent years. At German events, considerable attention is frequently devoted to polymers and composite materials - in a parallel way to the scope of the subject in this Delphi survey.

Who was surveyed?

In comparison with the other subject areas of the survey, the topic of those taking part from the university sector in the subject area of Chemistry and Materials is particularly high, but follows basic research in the field of materials development primarily at universities and other research institutes.

The topic which deals with *heat-resistant polymers at 450°C* was answered by 175 of those taking part in the survey and is accordingly top of the number of responses. At the bottom with only 63 respondents is the topic which discusses *data storage systems in the use of low-molecular information carriers*. The reason for the large differences in the numbers responding must surely be sought in the high degree of specialisation of the experts questioned. When it came to the first-mentioned topic, 46 % of experts with a low level of specialist knowledge responded, for the second topic the figure was 76 %. As an average of all topics in this subject area, the proportion of those persons who confirm they have a high level of specialist knowledge themselves - i.e. experts working in the field - is under 10 %. No significant differences can be determined with regard to the origin of the experts from industry or from universities.

Why is Chemistry and Materials an important field?

Compared with the other subject areas, outstanding importance in terms of economic development is accorded to the subject area of new chemical products and materials - by the representatives of the universities and industry in equal measure. Approximately 85 % of them refer to it (see fig. 4.4-1). In comparison with the other subject areas of the Delphi survey too, the subject field occupies a leading position in this respect. Because of that, the developments in Chemistry and Materials are accorded a key function for the German economy.

Figure 4.4-1:   What are the innovations of the subject field Chemistry and Materials important for?

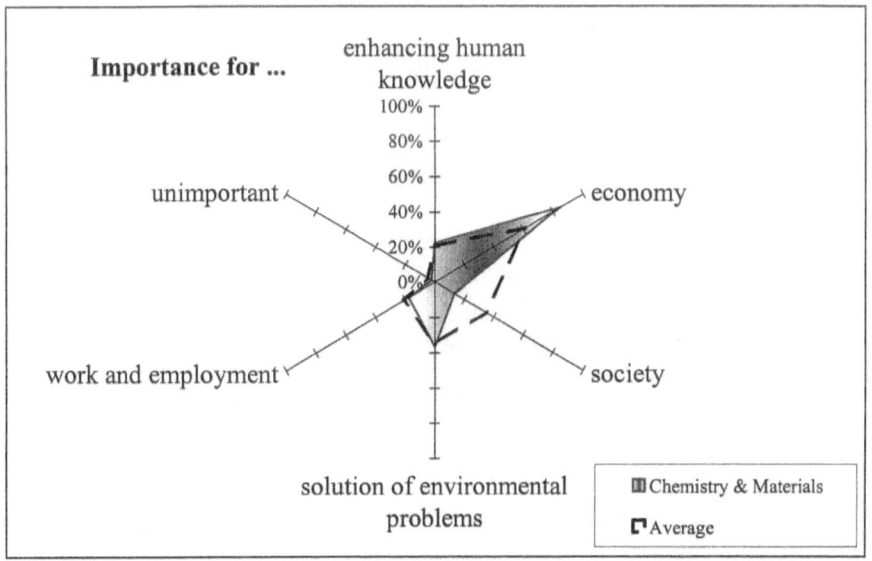

The importance for solving ecological problems is mentioned in second place by 35 % of the experts. This figure represents the average of all Delphi topics.

When will the topics become reality?

In the opinion of most experts only a few developments will be realised in the next 5 to 10 years: just five out of a total of 104. Most fulfilments will occur in the period of the next 10 to 15 years, namely 59 topics or 57 %. 40 developments are given a chance of fulfilment only after the year 2013. The developments in structural materials have far and away the greatest prospects of success.

Materials development and the introduction of new techniques in production patently require considerable staying power.

The German economy will and can only profit from these developments in the very long term. In addition, the very sophisticated developments described will not benefit the environment in the short- to medium term.

What is the level of Germany's research and development in Chemistry and Materials?

Despite the rather long-term prospects it would be inappropriate to detract from the status of this subject area. For on the one hand, it must be taken into account that the developments in this Delphi survey which are relevant to the present were deliberately not addressed. On the other, the long-term effect of research and development must not be overlooked. Germany is regarded as leading the world in 31 out of 104 topics, for more than half of which a realisation is not expected before the year 2011. For six of the topics, in fact, the realisation is only expected very much later.

What needs to be done for Chemistry and Materials of the future?

For the subject area of Chemistry and Materials, the measures stressed which could give the developments described more impetus can be found above all in

- strengthened international co-operation
- an increase in state and private support
- an improvement in the research infrastructure.

In particular, compared with the other subject areas, vehement demands are made for financial support. This is surely a reflection of the dominance of respondents from the university sector.

What follow-up problems might ensue?

Compared with other possible problems involving safety or in society, the resulting problems for the environment are considered particularly critical for the subject area of Chemistry and Materials. In the case of 31 topics, more than 80 % of respondents fear problems for the environment. However, at the same time these are the topics which in the opinion of the same experts can contribute to the solution of ecological problems. Solving ecological problems by the substitution of certain materials is therefore possible, but in future will perhaps cause other problems which cannot as yet be specified.

Key point: Can Germany as a chemicals location reconcile economy and environment?

The fact that economic and ecological perspectives do not have to be mutually exclusive is shown by the Delphi survey in convincing fashion for a series of topics in the specialist field of Chemistry and Materials. Below, those four topics are presented in greater depth in which more than 80 % of those surveyed point to their importance for economic development and for solving ecological problems as well as to Germany's leading role in research and development. Particular attention is paid to the additional comments of the experts.

The four topics presented here deal directly or indirectly with transport and traffic in general, in particular with vehicle construction. In this field especially, Germany is clearly accorded a leading role in research and development - above all in the lessening of negative effects on the environment. The fact that the latter cannot nevertheless be completely eliminated is documented by the fear expressed by many experts of resultant problems for the environment.

Powder technologies gain dominance in applications in the paint, adhesive and miscellaneous coating industry which are still dominated today by solvent-based or aqueous systems.

This goal can be achieved by the year 2006, i.e. in just eight years. This relatively short timespan is therefore regarded as quite likely, because the conversion process, to judge from the comments, has already been set in motion. On the other hand, a general substitution of conventional processes is thought to be unlikely, as an appropriate powder system would have to be developed for each coating problem and as a rule the coating powders cannot be stored for as long as one would like.

The polymer materials used in vehicle construction are produced on the base of less monomers and recycled again after use.

In the opinion of the experts, the recycling of polymer materials used in vehicle construction will gain general acceptance by the year 2007. Despite the positive aspects for the environment expressed in the survey, reference is made in the comments to the point that this development may be leading in a wrong direction: the comments indicate that the material utilisation of the synthetic materials employed is energy-intensive and that really for this application area energy utilisation is preferable to material use. The use of only one synthetic material for all functions in the car is not possible for technical reasons, because the requirements are too broad.

With special combustion engines made of ceramics it is possible to build a 2-litre car with low pollutant emissions without a catalytic converter.

In the opinion of the experts, a 2-litre car with a special combustion engine made of ceramics can be developed by the year 2014. In Germany, attention should be paid to strong Japanese competition - according to the group of experts surveyed here. The majority of them are also of the opinion that the development could be encouraged by appropriate laws. In the comments, reference is made in particular to the point that engines for 2-litre cars do not necessarily have to consist solely of ceramics. Furthermore, the engines will not be able to do without a catalytic converter. Finally, it is suggested that such small cars are presumably used too often as second or third cars and therefore a disservice would more likely be done to the environment as a result of their introduction.

Processes designed to produce fuels from microorganisms and algae are in common use, with the result that the proportion of these fuels (e.g. alcohol) reaches 10 % of world production.

Based on the survey, such bio-fuels should cover approximately 10 % of world production by 2020. However, the opinion of the experts is conflicting with regard to the possibilities of solving environmental problems in a comprehensive way using these bio-fuels. More than 91 % of those surveyed believe that there will be resulting problems for the environment - the highest proportion in this field. This is surely no accident, for why should bio-fuels produce less pollutants than conventional fuels? Using these fuels, the total CO_2 input into the atmosphere could certainly be somewhat reduced, but not altogether avoided. At the same time, the demands on mobility would have to change. Moreover, in the comments of the experts doubts are expressed that a scarcity of fossil fuels will become apparent as early as the year 2020 and that due to the rise in price of petrol and diesel caused by this the economic use of bio-fuels would be possible.

The selection of these four questions is designed especially to demonstrate in exemplary fashion the cross-sectional nature of the developments in the field of Chemistry and Materials. The variety and breadth of the questions on the one hand and the high level of participation in the response show that the chemical sector in Germany has a high status and has access to corresponding expertise. This becomes particularly clear in comparison with the Japanese Delphi surveys, in which chemicals only play a very subordinate role. Many future problems are connected with the fact that the materials conversion processes continue to be optimised. Considerable staying power is necessary for this, bearing in mind that technology has reached a high level thus far, but the efforts being made are economically as well as environmentally very important.

4.5   Health and Life Processes

(contributed by Bärbel Hüsing)

Structure of the innovation field

At the heart of the innovation field Health and Life Processes are the "classical" scientific disciplines of medicine and biology. Almost half of all the topics (49 out of 104) deal with aspects of medicine, namely prevention, diagnosis and treatment of human diseases; another 17 topics relate to the health care system. These key subjects of medicine and biology are linked by their common recourse to biotechnological methods and knowledge (e.g. genetic engineering and genomics): molecular medicine explains the cause and progress of diseases at the molecular level and, proceeding from there, develops causal treatment approaches, whereas the life sciences apply the common methodological toolbox to identify the molecular basis of life processes (covering a total of 19 topics). It is biotechnology which exploits these life processes for technical systems. Nineteen topics concern selected areas of biotechnology.

Who was surveyed?

Compared with other innovation fields, the topics relating to the clarification of underlying phenomena or to the development of new technology claim a disproportionately large amount of attention. Only about one-third of the topics relate to application. This means that the situation in this subject area is the exact opposite of the one which prevails in the average field of the survey, reflecting the fact that innovations in medico-pharmaceutical research and in biotechnology are strongly science-based. The percentage of specialists from universities is correspondingly high (about 45 %), while those operating in the business world are considerably fewer in number (accounting for about 26 % of the sample) than is the case in other innovation fields.

Why is Health and Life Processes an important field?

The study of health and life processes is one of the main keys to the social development and the advancement of human knowledge. The "importance quotients" for these areas are far higher than the average for all other innovation fields, while very little is expected from them as a source of economic development, work and employment or as a means of solving ecological problems (fig. 4.5-1). Experts from science and business largely agree in this assessment, except that the industrial experts are slightly more optimistic than their academic counterparts about the economic potential and slightly less optimistic about the knowledge that the study of health and life processes might generate.

Figure 4.5-1:   What are the innovations in the field of Health and Life Processes important for?

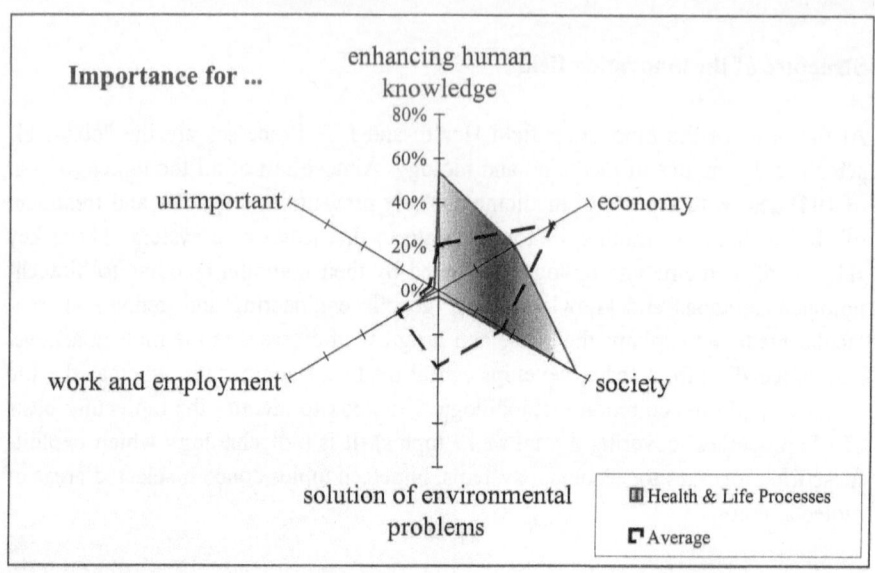

One reason why these innovations are so important to the development of the society is that they help to prevent and eradicate diseases and to the quality of life. Top priority is given to research into diseases which are incurable or which can only be treated symptomatically. These include cardiovascular diseases, metabolic disorders (e.g. diabetes and gout), cancer, neuro-degenerative diseases (e.g. Alzheimer's and Parkinson's disease), degenerative diseases of the locomotor system, immunodeficiencies and "new" infectious diseases such as AIDS, Creutzfeldt-Jakob disease and hepatitis C.

Another reason why the experts regard this subject area as an extraordinarily important social development factor is that new knowledge about the nature of illnesses and new means of diagnosing and treating them touch upon focal points of human life and "being human" and will therefore have diverse and sometimes profound moral, cultural and social repercussions. Analysis of the comments on the various topics highlighted the following main problem areas:

The topics concerning the beginning and end of human life invariably provoked a stream of comments - statements such as *prophylactic measures to prevent congenital defects*, the *"morning after" pill*, the *broad use of methods to limit population growth, euthanasia on demand for elderly and seriously disabled patients*, the *artificial prolongation of life in intensive care units* or by *conservation of living organisms on the basis of the hibernation principle*, or *meeting the demand for donor organs by means of public education and instruction*.

Another socially important problem area relates to the question of human dignity, in other words the extent to which the individual is, will be or should be the subject or the object of medicine and to which dealings with patients are, will be or should be customised or standardised. Examples of relevant topics are *robots to assist in the care of the sick, the elderly and persons with serious physical disabilities or mental disorders*, the *use of remote surgery systems with virtual reality* and even *self-reproducing technological systems* or the *specification of standard schemes of treatment by health insurance providers*. The large number and wide diversity of the comments indicates that this is an extremely complex set of issues to which there are no simple uniform answers; on the contrary, a highly discriminating approach is required. This implies the need for a broad debate within society on the scientific and technological developments highlighted by this survey.

An expansion of human knowledge is primarily expected to materialise in the neurosciences, in areas such as *experimental techniques for the simultaneous observation and analysis of a number of neurons, clarification of the mechanisms governing the formation of neural networks* and *elucidation of the neurobiological bases of cerebral functions*. The experts also believe that the *complete sequencing of the human genome* will serve primarily to provide knowledge; they consider the Human Genome Project carried out in worldwide co-operation to be distinctly less significant in terms of its potential impact on economic and social development.

Interestingly, the *widespread use of biological production by cell-free synthesis* and the *production of tailor-made enzymes for widespread use in all areas of material conversion* are among the developments to which the respondents attach great importance (relative to other developments in this field) in terms of economic development potential and as sources of solutions to ecological problems. *More flexible forms of co-operation between industry and university research facilities* were also considered to be very important as potential sources of economic development.

When will these topics become reality?

Because of the sharp focus on basic research in the topics of this innovation field, the experts expect that it will take some time for them to become reality. The average estimate is around the year 2012. Solutions with practical applicability are therefore expected to emerge later than in most nnovation fields areas of the Delphi survey. Interestingly, issues relating to cures for diseases are not only considered extremely relevant to social development but also appear in the eyes of the experts to be altogether achievable, although they expect it to take 10 to 15 years, and even longer in some cases, until such cures materialise. The view of the experts is that these targets can primarily be achieved by medical research that is rooted in scientifically verifiable knowledge.

Little significance is attached, on the other hand, to alternative forms of therapy. About a quarter of the experts regard these as unimportant, and the same number again consider that the *integration of methods currently regarded as alternative into orthodox medicine by establishing chairs in such disciplines at German universities* and the *widespread use of natural substances such as those used in Chinese medicine* can never be achieved. This does not exclude the possibility that treatments from the field of alternative medicine could be integrated into orthodox medicine if evidence of their effectiveness is provided. A large number of experts also regarded *self-reproducing technological systems* and the *conservation of living organisms on the basis of the hibernation principle* as both unimportant and unachievable.

What is the level of Germany's research and development in Health and Life Processes?

Compared with its general position for all innovation fields, German research is rather poorly placed in the area of Health and Life Processes, where the leading position of the United States is uncontested (fig. 4.5-2). This assessment, which may seem surprising in view of the traditional strength of the German pharmaceutical industry, may be based on the fact that multinational companies prefer to cooperate and invest in the United States, especially in the field of biomedical research. A host of reasons are cited for this; besides factors relating to market access and proximity to the licensing authorities, the respondents also referred to the more favourable structural conditions in the U.S.A. for the transfer of knowledge between institutions engaged in basic research and clinics.

Efforts to eliminate structural deficits in Germany, for example *fundamental structural reforms in German university hospitals with a view to creating better conditions for clinical research*, are regarded by the majority of the experts as an important contribution to social and economic development. However, from the experts' comments, which reveal that the implementation of such reforms is not expected before the year 2008 and that a relatively high proportion (7.7 %) of the experts do not believe that such an aim can ever be achieved, it is evident that the "staying power" of the present structures is quite formidable. Compared with other fields, Health and Life Processes is an area in which Japanese research is generally rather poorly placed, although it is strong in the areas of diagnostics and biotechnology. German and Japanese experts largely concur in this assessment.

Figure 4.5-2:   Germany's R&D position compared internationally

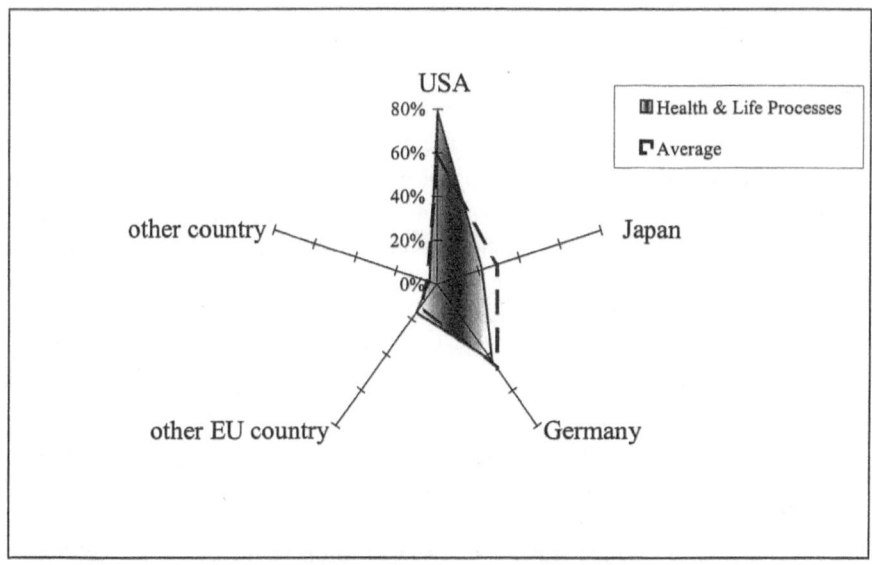

What needs to be done for Health and Life Processes of the future?

Given the sharp focus on basic research and the complexity of the questions addressed by research, it is understandable that international co-operation, infrastructure improvement and funding by third parties are seen as key instruments in the bid
to achieve the established targets. The importance of international co-operation is
not only recognised for resource-intensive major projects such as the *full sequencing of the human genome* or *the elucidation of the protein function from the gene
sequence* or for cross-border projects such as the *establishment of international epidemiological databases*.

The explicit call for international co-operation, for instance in the *development of
effective AIDS treatment* or in the *development of cancer treatments by influencing
signal transmission during the development of cancerous cells*, shows that relevant
knowledge is generated nowadays on an international scale and through networks,
in which the widest diversity of players, from universities and non-university research institutes to manufacturing companies, play a part. This means that researchers have to pursue an active policy of establishing close links with the leading international centres of excellence from which knowledge is generated and of integrating themselves into the appropriate networks. This implies a need for international
innovation management to locate and acquire the requisite know-how. Exchanges
of personnel between business and science are considered more important by the
academic specialists in the sample than by the business experts.

External support is regarded as an appropriate measure for various forms of highly basic research. The principal purpose of such research is to push back the frontiers of human knowledge without there being any obvious significant economic relevance. It includes such topics as *experimental techniques for the simultaneous observation and analysis of a number of neurons* and *elucidation of the mechanisms governing the formation of neural networks*. The experts also call for greater financial support for developments which are still more or less in their infancy, but which are regarded as particularly significant in economic terms. These would be developments in which the experts postulate the potential of an entirely new branch of technology, comparable with genetic engineering or rational molecular design, etc. The relevant issues are the *use of processes for the identification of new molecular functions or for the optimisation of known molecular functions, based on evolutionary mechanisms*. Precisely because the economic relevance of this technology has already become evident in the early stages of its development there is a danger that, because of unduly rapid commercialisation, fundamental issues will not be addressed to the extent that they ought to be if the basic concept is to be refined and hence the full potential of evolutionary processes exploited.

Finally, third parties should undertake to support projects involving research which the experts consider to be ecologically or economically important, but for which the cost of research is disproportionately high in relation to its subsequent value to the private sector. Such research includes the *production of tailor-made enzymes for widespread use in all areas of material conversion* or the *clarification of the neurochemical and genetic causes of alcoholism*.

What follow-up problems might ensue?

We have already referred above to the exceptionally high number of moral and socio-cultural problems which might arise from developments in this innovation field. Scientific and technological developments in the area of Health and Life Processes are hardly likely to create ecological problems, but major safety problems are feared in some disciplines. On the one hand, the experts are concerned about developments such as the *use of remote surgery systems with virtual reality*, the use of *health-online systems to increase self-treatment by patients*, the *linking of at-risk patients with their GPs by means of diagnostic early-warning systems and monitoring software* or a *market-based system of health care, in which the prices charged by the various service providers are determined by supply and demand*, since they fear that the present high quality standards of the German health care system could not be maintained if such developments took place.

Safety problems that mainly consist of unwanted side-effects are anticipated by the experts in topics such as the *use of radiation sensitisers in cancer treatment*, the *widespread clinical use of somatic gene therapy* and the *use of artificial blood*. Problems of data protection, insurance cover, accessibility of data and the "right not to know" arise in connection with the *use of genetic diagnostic processes with which a higher level of individual susceptibility to a particular disease can be identified*, with the *establishment of universally accessible international epidemiological databases* and with a *German cancer register*.

Key point: The introduction of new techniques will change the principle of solidarity that underlies the German health care system

The health care system is the social, institutional and organisational framework within which medical services are provided and medical research is conducted. In the industrialised countries, health care is a value in its own right, a facility for which people are prepared to spend money and which is not subject to the usual cost-benefit considerations. However, in view of the fact that while public funds become tighter, health expenditure continues to rise, generally at a faster rate than GNP, cost-consciousness is gaining ground in almost every industrialised country. New measures are being taken to cut costs and to increase efficiency.

The expert respondents, however, give a clear thumbs-down to the idea of a *single federal health insurance scheme with a catalogue of services*, which almost half of them consider to be impracticable. On the other hand, only 13 % of the experts regard a *market-based system of health care, in which the prices charged by the various service providers are determined by supply and demand* as impracticable. In other words, most of the experts think that a market-based system of health care is likely to come about but consider such a step to be tantamount to the introduction of a two-tiered system of medical care and thus to be socially regressive. *Microsolidarity-communities in the form of mutual societies, providing care in the event of sickness, infirmity and destitution* are also regarded as unworkable by almost one-third of the experts, because such bodies would not be a suitable means of ensuring blanket coverage. Any attempt to erode the solidarity principle, for example by introducing *higher health insurance premiums for those in whom genetic diagnostic processes have identified a higher level of susceptibility to a particular disease* must be prevented by the creation of appropriate statutory provisions, according to 85 % of the experts.

4.6 Agriculture and Food

(contributed by Bärbel Hüsing)

Structure of the innovation field

The field of Agriculture and Food covers agriculture, forestry and fisheries on the one hand (37 topics), activities which are closely associated with the use of resources such as soil, water and genetic resources. Accordingly, 29 topics relate to the sustainable use of resources or to the protection of nature and the natural environment. The use of remote-sensing systems and information technology for purposes such as the observation of changes, resource management, regional planning or information and prediction services are the subject of eight topics.

On the other hand, the innovation field contains 28 topics on the processing and consumption of foodstuffs. Biotechnology has traditionally played an important part in food production and processing. New biotechnological processes and genetic engineering applications are now ready to be put into practice - and some are even ready for marketing - in plant and livestock breeding, agricultural production and food processing as well as modification of the health effects of nutrition. This explains why at least 35 of the total of 101 topics relate to the use of biotechnology and genetic engineering in agriculture and food production.

Given that both agriculture and food are areas where research activity is relatively thinly spread, it comes as something of a surprise to discover that more than 40 % of the topics in this subject area relate to the clarification of basic phenomena or the development of initial prototypes, while only slightly more than 50 % of the topics concern the practical application and the broad dissemination of innovations. This puts Agriculture and Food in third place among the subject areas of this survey, behind the highly science-based subject areas of Big Science Experiments and of Health and Life Processes. The reason could lie in the great importance of biotechnology as a strongly fundamentally-oriented enabling technology for this subject area and in the fact that numerous biotechnological developments have only just reached the prototype stage.

Who was surveyed?

About 30 % of the experts work in universities, while a further 30 % are from industry. Compared to the other innovation fields, a higher proportion of the experts (20 %) are civil servants, working for bodies such as the federal research institutes. Experts from universities and other institutional establishments more often rank their expertise higher than the respondents from companies and the public service.

Why is Agriculture and Food an important field?

The question why research into agriculture and food is important elicits answers which are little different to those for the average innovation field. The importance of R&D in this area is principally considered to lie in economic development, social development and in solving ecological problems, whereas it is not expected to contribute to any particularly great extent to the advancement of human knowledge or to work and employment (fig. 4.6-1).

Figure 4.6-1:   What are the innovations in the field of Agriculture and Food important for?

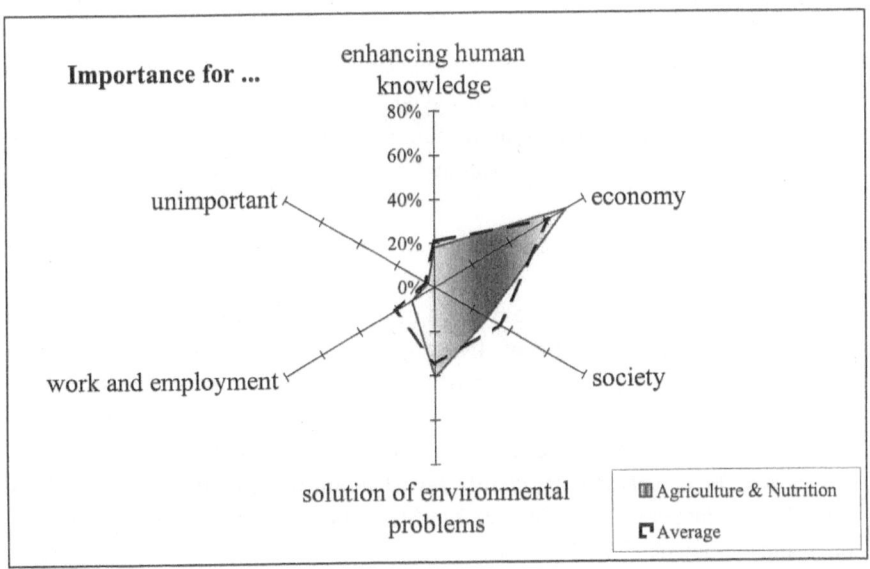

The importance of this innovation field in the context of economic development is reflected in the fact that 34 of the 101 topics are regarded by more than 90 % of the expert respondents as economically important. It is interesting to note that four of the five topics with the highest importance quotient in terms of economic development relate to new biotechnological processes (*transgenic plants with improved compositions; feed additives produced with the aid of genetic engineering; production of high-value substances with cell cultures in large-scale bio-reactors; selective modification of crop yields*). The fifth topic relates to the use of modern food technology in general (*successful market launch of numerous innovative products by small and medium-sized food producers, using the latest food technology*).

This shows that the food and beverage industry is facing the major problem of a generally stagnating market, where growth is only occurring in specific segments and is usually accompanied by falling sales in others. People in the industrialised countries will not eat and drink more in the future than they do today, but diets and consumption patterns will change. In the context of intense competition among food producers, strategies consist of increasing turnover by developing new products or product ranges or in making efficiency gains in the production process to cut production costs. Modern food technologies play a leading role in both of these strategies. Moreover, about half of the experts regard such technologies as important for work and employment.

With a view to solving ecological problems, more than 90 % of the experts attach importance to research into the sustainable use and recultivation of (tropical) forests and methods to reduce emissions (e.g. nitrogen, pesticides) in agriculture and aquaculture. Contributions to social development are primarily expected to come from innovations relating to the impact of foodstuffs on consumers' health. These include the *causal investigation of the influence of food on human health* and of the *influence on human health of certain methods used in food production and processing*, as well as the dissemination of nutritional advice among *consumers, to enable them to act accordingly* and *clean water supplies for all people.*

When will these topics become reality?

Although many topics relate to questions of basic research, the experts assume an average time of realisation of some twelve years (2010). This puts Agriculture and Food about the middle of the range of innovation fields covered by the survey. However, an above-average number of innovations are regarded as achievable by the year 2005. Only three visions are consigned to the period beyond 2020. In this subject area, then, a particularly well-balanced choice of topics between some rather unsurprising and generally predictable developments of the present body of research and some unrealistic Utopian dreams has been made.

What is the level of Germany's research and development in Agriculture and Food?

German research and development in the area of Agriculture and Food is roughly on a par with the strength of its R&D in most other innovation fields of the survey. It is considered to be especially strong in the use of resources, the processing and marketing of foodstuffs and in the area of environmental impact, where the experts even put Germany ahead of the United States. American researchers head the field in all other aspects of agriculture and food, with a particularly commanding lead in breeding, information technology and agricultural production. Japanese research, although generally classed as comparatively weak in this area, has relative strengths in the areas of food composition and nutritional practice, and in food processing and marketing (fig. 4.6-2).

Figure 4.6-2:   Germany's R&D position compared internationally (sub-groups)

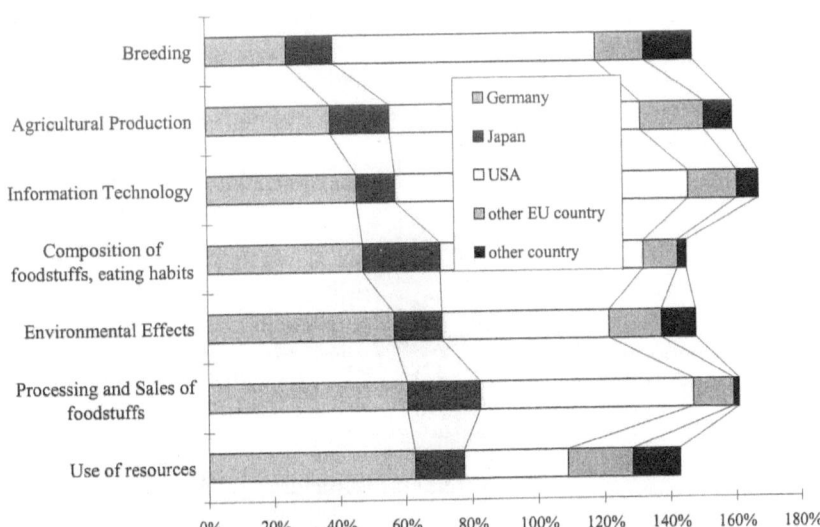

What needs to be done for Agriculture and Food in the future?

The measures identified by the experts as necessary and appropriate ways to improve the present situation are not essentially different to those prescribed in most of the other innovation fields of the survey (fig. 4.6-3). A slightly higher priority is attached to international co-operation, and the R&D infrastructure is ranked slightly lower. Interestingly, however, the experts specifically allocate particularly important measures to individual sub-categories in the area of Agriculture and Food.

It emerges from numerous comments that when the experts call for better training they are thinking primarily of the instruction and education of consumers. This applies especially to knowledge of healthy eating habits and of nutrition-related diseases. Consumer education is therefore regarded by the experts as the main means of *halving the present number of cases of infection caused by salmonella or other enterobacteria* and would make it unnecessary to develop *measuring instruments for household use with which the freshness of food and its degree of contamination by microorganisms can be determined in a matter of seconds.* Better training is also regarded as a suitable alternative to *halving the number of unskilled employees in the food processing industry by means of process automation and computerised control,* a development which the experts expect by the year 2006.

Figure 4.6-3:   What measures should be taken?

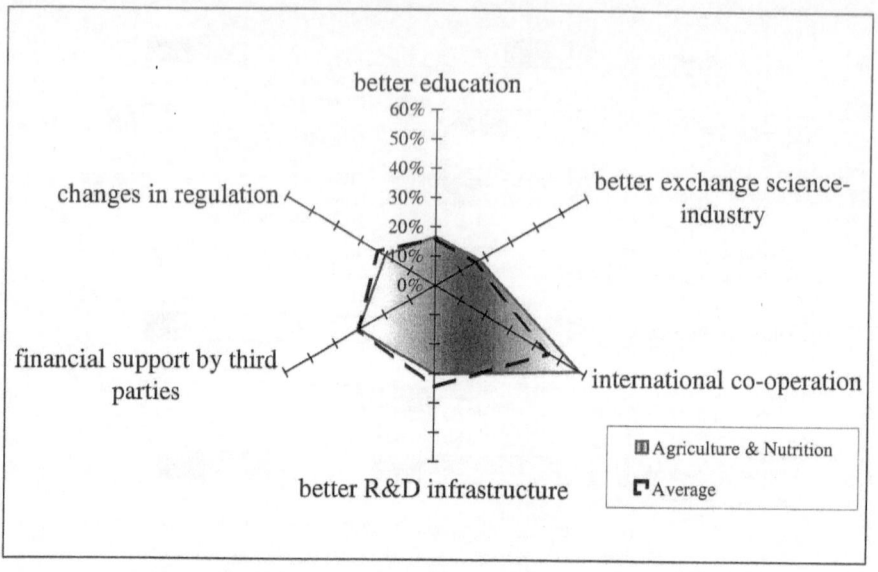

For the acquisition of know-how on the *use of the latest food technology with which small and medium-sized food producers can introduce numerous innovative products into the market*, almost half of the experts cite the importance of improved training. But almost three-quarters regard staff exchanges between business and science as the more important measure in this instance. At the same time, the experts are sceptical as to whether small and medium-sized businesses can actually acquire such know-how. And no fewer than 16 % of the experts consider that this aim could never be achieved, not least because of the costs involved.

It is understandable that a relatively large number of specialists restrict their calls for more staff exchanges to the topics concerning food, since much of the food industry, consisting chiefly of small and medium-sized companies, can afford little or nothing in the way of research facilities for lack of financial and human resources. The same situation is also reflected in the fact that experts from industry attach higher priority to exchanges between business and science than do experts from the scientific world. Nevertheless, if we analyse the statements in relation to which exchanges of staff are considered important (*equipment for food processing that only has to be washed half as often as hitherto; large-scale production of high-value substances with cell cultures in bio-reactors; clarification of the causal relationship between the structure of food ingredients and their technical properties*), it emerges that international co-operation and the R&D infrastructure are considered even more important to each of these individual innovations.

Particular importance is ascribed to the improvement of the R&D infrastructure in the area of food technology. More than half of the experts see the need for a better infrastructure in connection with the development of processes that can be used to measure the quality of foodstuffs during product development, in food processing, and in the household.

International co-operation, on the other hand, is primarily considered to be vitally important when it comes to the use of resources and the assessment and management of environmental impact. More than 90 % of the experts regard international co-operation as one of the main prerequisites for the sustainable use of global resources - especially the rainforests - and for the breeding of crop plants which can even be grown in arid climates.

An unequivocal answer is given to the question concerning the field, in which support by a third party is most important: in the experts' view, it is the use of renewable raw materials for industrial or energy production. However, although much depends on the prevailing political and economic conditions, it is estimated that a period of about 15 years is needed before the percentage of renewable resources will have increased significantly. But at the same time, the experts consider it possible that the global food situation will make it necessary to devote almost all productive farmland to food and fodder crops.

As well as support funds, international co-operation in the use of renewable resources is also very important. More than 90 % of the experts believe that R&D in this area is one of the keys to economic development, since it can open up additional or alternative sources of income to farmers. About 60 % see it as a means of solving ecological problems.

Innovations designed to minimise environmental pollution, such as attempts to reduce nitrogen emissions or the use of pesticides in agricultural production, are also heavily reliant on financial support. These, however, are largely developments that can only be realised in the distant future, if at all.

What follow-up problems might ensue?

Together with energy, the environment and chemistry, Agriculture and Food is one of the innovation fields of the survey in which the most serious environmental impacts are expected. The subfields environmental impact, use of resources, breeding, and agricultural production are the main contributers.

As far as safety problems are concerned, Agriculture and Food ranks somewhere in the middle of all innovation fields covered by the survey. The experts see hygiene and the wholesomeness and the nutritional value of foodstuffs as the main areas in which safety problems can arise.

The experts anticipate considerable problems in the social and socio-cultural areas. Only in the subject areas of "bio-medicine" and of "services and information" are there more frequent references to the possibility of far-reaching effects on society. The specialists highlight three main problem areas: the first concerns the impact of food on human health, the second concerns food and nutrition as a major cultural asset rather than merely a vital necessity, and the third relates to the use of genetic engineering in the agro-food sector, a development which is largely opposed by consumers (for details see an additional, more detailed Delphi study, Menrad et al. 1998).

Key point: Food produced with the aid of genetic engineering will become an integral feature of the German market

The logic, necessity, extent and objectives of the use of genetic engineering in the agro-food sector are the subject of intense debate in Germany. At the time of the Delphi survey, genetically-modified soy beans had become the first genetically engineered food product to appear on the German market. How do the experts involved in the survey expect the situation to develop? *Successful forms of intervention (de-listing of food items, boycotts) which make the trade organisations compel food producers and processors not to use genetic engineering processes and products created with the aid of genetic engineering,* while they may be relevant for the next few years, tend to be regarded by the experts as a temporary phenomenon. They will not be able to halt the long-term development leading to the use of genetically-modified products and genetic engineering processes.

The main users of this technology will be multinational companies, which will only reinforce the trend towards industrial concentration that has been in evidence for some time. The topic that *foodstuffs produced wholly or partly with the aid of genetic engineering will account in the coming decades for 30 % or more of all food sales in Germany* is considered feasible by 95 % of the experts, provided that consumers are fully informed, that their wish to have genetically-modified products labelled is respected and that there are no adverse experiences with such products. A *tightening of the food labelling requirement, making it compulsory to give details of every ingredient and technological process used in the production of all processed foodstuffs and beverages,* however, is regarded by more than one-third of the experts as impracticable, since such a wealth of information would be too much for shoppers to digest. The experts differed over the extent to which research was needed *to clarify the influence of certain methods used in food production and processing (e.g. genetic engineering, radiation and microwave heating) on human health (e.g. allergies).*

4.7 Environment and Nature

(contributed by Harald Hiessl)

Structure of the innovation field

The subject area of Environment and Nature comprises a total of 76 topics, covering a very broad spectrum of issues relating to environmental protection and nature conservation. To make the subject area more easily surveyable and to facilitate interpretation of the survey results, the statements were divided into eight groups: atmospheric protection (13 topics), habitat protection and forestry (5 topics), the urban environment (11 topics), waste management (5 topics), soil protection (8 topics), protection of hydrological systems and management of water resources (17 topics), marine protection (4 topics) and environmental monitoring and information (13 topics).

Compared with other innovation fields, Environment and Nature has a particularly large percentage (over 50 %) of topics relating to the "broad dissemination/ general application" of the stage of the innovation process. This shows that in the area of Environment and Nature the practical application of new technology plays a relatively large role in comparison with the development of new products or basic research.

Who was surveyed?

The percentage of those surveyed who state that they possess a great deal of specialist knowledge is about average. Compared with other subject areas, the respondents with a high level of expertise are more evenly spread among the various sectors of the national economy. One reason for this is no doubt to be found in the very broad spectrum of topics dealt with in this section. For all that, Environment and Nature resembles other innovation fields in that the majority of the respondents with a high level of specialist knowledge are from private non-profitmaking research institutes and from universities, closely followed by respondents from the civil service. On the other hand, the percentage of respondents from industry and other sectors of the economy is distinctly lower. One of the likely reasons for this is that the purpose of environmental research still tends to be seen in the development of technology to protect the environment or in the monitoring and control of pollution rather than as a potential economic factor in terms of greater efficiency in the use of resources.

Why is Environment and Nature an important field?

As expected, the main thrust of the topics in the subject area of Environment and Nature, in the eyes of an absolutely overwhelming majority of the respondents, lies in solving ecological problems (fig. 4.7-1). The steps towards such solutions that are presented in the topics are considered to be distinctly more important in terms of economic development than in their potential contribution to the solution of social problems. One reason for this lies in the fact that environmental protection is increasingly being incorporated into the production process in the form of integrated environmental protection technology, generating a high level of know-how and of added value.

Figure 4.7-1:   What are the innovations in the field of Environment and Nature important for?

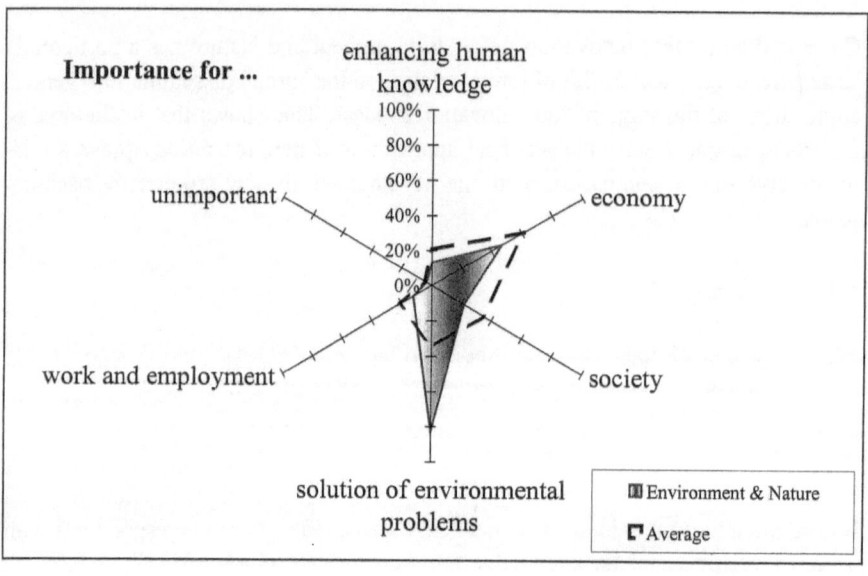

An interesting aspect is that the experts from the worlds of science and industry attach roughly the same importance to the economic and the ecological problem-solving potential of these topics and that both groups of experts cite the solution of ecological problems almost twice as often as they cite economic development. So it is evident that the expert respondents do not think of environmental technology as a "brake on the economy".

The following specific topics emerge as the most important in the various spheres of Environment and Nature:

Research on *atmospheric processes and climatic models, climate alteration* and *earthquake prediction*, along with other topics relating to the alteration of climatic conditions, are especially important tasks to advance human knowledge.

Technology that can be used on a decentralised ("on-site") basis to purify drinking water, the biotechnological development of salt-resistant, aridity-resistant and high-yield plant varieties, marine farming, appropriate fertilisation and *remote-sensing systems for flood prediction* are of particularly great importance to economic development.

New settlement structures designed to reduce traffic-induced pollution, restriction of land use, nature conservation areas in Germany and *flood prediction methods* are very important matters in terms of social development.

Significant contributions to the solution of ecological problems are seen in *remote monitoring of the seas to combat pollutant dumping, lowering emissions of persistent pollutants to environmentally acceptable levels, halting the large-scale clearance of rainforests, finding and introducing substitutes for CFCs and halon gases* and *biotechnological and physical processes for the in-situ cleaning of polluted groundwater.*

Particular importance in terms of business activity and employment is ascribed by the experts to *ecosystem-friendly farming in Germany*, the *recultivation of damaged rainforests, desert recultivation techniques, vehicles with low nitrogen oxide levels* and *soil-protective farming methods and agricultural engineering processes.*

As we interpret these findings, it is worth noting that in every category the experts attach special importance not only to topics of greater local relevance but also to those which relate to global issues. So there is clearly no tendency among the specialist community to "sweep global problems under the carpet".

When will these topics become reality?

With regard to realisation times, there are no significant differences between the average dates cited in the Delphi survey in general and those estimated for the innovation field of Environment and Nature. What are interesting, however, are the ranges of dispersion for the realisation times in the individual environmental areas. They reveal that topics relating to supply and disposal problems, in other words to more "local" issues which are directly experienced by every individual, are assumed on average to be realisable within a considerably shorter time than topics focusing chiefly on "global" issues, which tend to lie outside the experts' direct experience. The median values for the times of realisation estimated for innovations in the various groups of topics are shown in table 4.7-1.

Table 4.7-1:   Average estimated times of realisation in certain environmental
                areas:

Refuse disposal:	2009 ± 3
Protection of hydro-logical systems and management of water resources:	2010 ± 4
Atmospheric protection:	2013 ± 4
Habitat protection and forestry:	2013 ± 4
Environmental monitoring and information:	2013 ± 4
Marine protection:	2014 ± 4
Soil protection:	2014 ± 5
Urban environment:	2016 ± 5

Foremost importance in terms of economic development is assigned by the experts to refuse disposal and to the protection of hydrological systems and management of water resources. Moreover, German R&D in these areas is considered to be well ahead of the field. This suggests that appropriate new technology should be introduced quickly and judiciously in these fields to benefit both the economy and the environment and to build on the strong position of German environmental technology in the international market.

What is the level of Germany's research and development ?

Looking at the experts' assessment of the R&D position, we see that Germany features most often as one of the leaders in the field, a status assigned to it by 65 % of the experts; the United States follows with 45 %, then come Japan (25 %) and other EU member countries (15 %). A more refined picture of the German experts' appraisal of their country's R&D position in relation to other countries emerges when the various groups of topics are considered separately (fig. 4.7-2).

Figure 4.7-2 shows that the experts involved in the survey only assign a leading role to Germany in the groups of topics relating to "biotope protection / forestry", the "urban environment", "waste-processing", "soil protection" and "protection of hydrological systems and management of water resources", while they acknowledge the leading position of the United States in the areas of "atmospheric protection", "marine protection" and "environmental monitoring and information". The German experts put Japan ahead of their own country's R&D in respect of only one group of topics, namely those under the heading of "marine protection", where Japan is seen to have a distinct edge over Germany, which is not surprising in view of Japan's island situation.

Figure 4.7-2:   Germany's R&D position compared internationally (sub-groups)

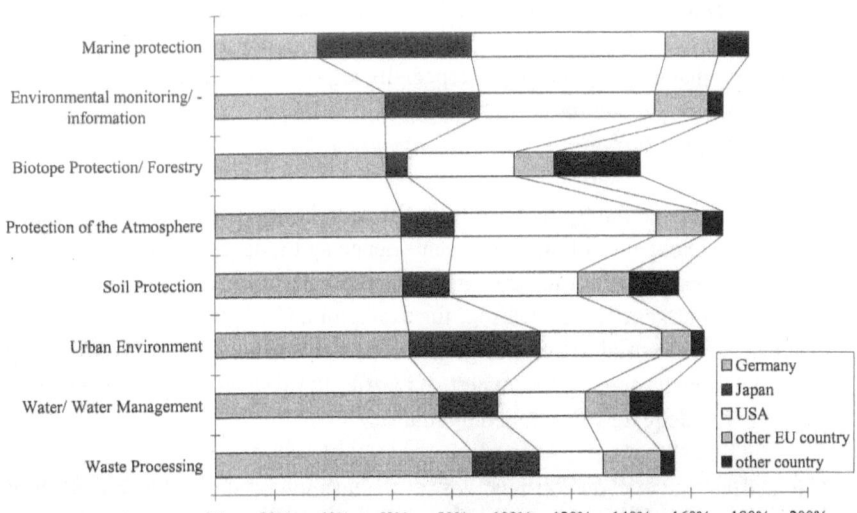

What needs to be done for Environment and Nature of the future?

What political measures are required in the area of Environment and Nature? The respondents regard international co-operation as the main means of implementing the innovations described in the topics within this subject area. This result is in line with the findings in most other areas of the Delphi survey. However, although improvement of the R&D infrastructure comes a clear second in most of the other subject areas, in the area of Environment and Nature changing the regulatory framework ranks as highly as international co-operation as a means of implementing the innovations. This not only indicates a clear need to act on the regulatory framework governing environmental matters but also reveals a demand for appropriate changes in the law as a necessary prerequisite of improvements in the environmental situation.

If we distinguish between individual groups of topics, it emerges that the experts see regulatory changes as the preferred option for statements relating primarily to "local" environmental problems, while they regard international co-operation as the key to the successful implementation of the topics relating to the more "globally"-oriented aspects of the environment. What is especially noticeable is the high percentage of respondents who call for regulatory changes ("re-regulation" is the recurring word) as an important means of achieving substantive improvements in the realm of waste management. There are obviously new avenues to be explored here.

A particularly large number of experts call for improved international co-operation in the fields of *remote sensing for flood prediction*, *stratosphere monitoring* and *remote sensing to monitor marine pollution and ocean currents*.

Regulatory changes are considered especially important in connection with the *definition of water protection areas, the restriction of land use* and *(artificial) rainwater infiltration (urban rainwater management)*.

Lack of training is chiefly referred to in the context of environmentally-compatible agriculture. Exchanges of staff between science and industry are regarded as a key to the development of more ecologically acceptable biocides (with the aid of biotechnology and genetic engineering, for example) and to physical processes for the treatment of industrial sewage (e.g. *membrane processes, aerogels, plasma processes*). Obviously, science has expertise to offer in these areas, expertise which has not yet been adequately developed by industry.

The fact that the experts consider staff exchanges important in the context of modernising the technology used for water supply and sewage disposal indicates that they regard more extensive integration of strategies and experience from the private sector as an opportunity to ease the technological "modernisation log jam". The calls for the *protection of water* as the most important necessity for life, the *reduction of land use* and the *cleaning-up of hydrological systems impaired by extensive sealing of soil* should encourage the legislature to enact the necessary changes in the statutory framework.

What follow-up problems might ensue?

A disproportionately large number of experts from the area of Environment and Nature expect environmental problems to result from innovations in their field. This seems paradoxical at first sight, but the experts' responses are merely underlining the environmental relevance of most of the topics advanced in this subject area. On closer inspection, it emerges that their main fears relate to problems that might ensue from the management of marine resources in the form of *marine farming* and the *cultivation of microorganisms*. Other environmental problems might result from the use of processes *whereby water sediments can be cleaned in situ of heavy metals and organic pollutants*. Any safety or social problems that might ensue in this field are considered insignificant by the experts.

> **Key point: Germany can hold its lead in the realm of environmental technology**
>
> Most of the experts consider Germany to be the world leader in R&D in the field of Environment and Nature. German R&D is believed to be especially strong in the areas of *"waste processing"* and *"protection of hydrological systems and management of water resource*s". The experts not only assume far shorter times of realisation periods for the topics in these groups than for all other groups of topics (table 4.7-1), they also attach the greatest importance to them in terms of economic development potential. This serves to indicate that the judicious introduction of new technology in these two areas not only offers scope for ecological improvements but can also consolidate Germany's competitive position in the international market for environmental technology.
>
> But: a clear majority of the respondents call for changes to the regulatory framework in the areas of *"waste processing"* and *"protection of hydrological systems and management of water resources"*, as well as with regard to the *"urban environment"*, as the foremost requirement in this field ("re-regulation" is the recurring word). The high standing of German R&D and its competitive edge in the world market can only be maintained if the regulations are amended. The experts believe that the technological and institutional log jam that is still delaying the modernisation of water supply management could be eased if experiences and strategies from the private sector were incorporated to a greater extent into the system.

4.8   Energy and Resources

(contributed by Harald Bradke)

Structure of the innovation field

The nub of the subject area of Energy and Resources lies in the production and conversion of primary energy sources (47 topics, including 18 on renewable sources of energy, ten on nuclear energy and nine on the subject of hydrogen/fuel cells), followed by 36 topics relating to the efficient use of energy, including eleven topics on transport and the same number again on the industrial use of energy. The transport and distribution of energy (9 topics) and energy storage (4 topics) complete the energy section. In addition, the survey covered 18 topics on non-renewable raw materials, with the main emphasis on recycling and avoidance of waste.

This subject area is dominated by topics relating to the economic development of technology and the development of industrial applications. Purely technical developments and basic scientific research play a minor role. Nevertheless, the presumed realisation dates for the topics in this field lie in the distant future, which is explained by the relatively long development periods and reinvestment cycles required for the various forms of energy technology described in the topics.

Who was surveyed?

The number of responses to each topic lies between about 70 and 170, the average being 123 responses. The figures are fairly stable from one group of statements to the next, with only slightly fewer responses to the topics on efficient energy use (118 per topics on average). Some 40 % of the respondents, which is above the average for the study, come from businesses in the private sector, which is well in line with the more application-based nature of the topics. Slightly fewer than 30 % of the replies (below the average for the Delphi study) are from experts working in universities. The remainder are divided in roughly equal measure between private non-profitmaking research institutes, the civil service and "others".

The experts' assessment of their own specialist competence puts them among the top-rated groups in the study. At the present time, more than 15 % of the respondents from both industry and the private non-profitmaking research institutes are actively involved in work on the questions raised in the various areas of this field. Active research into these questions is being conducted by almost 12 % of the respondents from industry but only by about 7 % of the scientists from the university sector. An above-average proportion of the "active" experts are involved in work relating to the topics on renewable sources of energy and on fossile-fired power stations (18 % compared with 12 % for all aspects of the subject area), while conspicuously few respondents are actively involved in the areas of exhaustible resources (7 %), transport and efficient energy use in industry (6 % for each).

Why is Energy and Resources an important field?

In view of the great importance attached to energy supply within the national economy and the ecological problems associated with the conversion and use of energy, it is no wonder that 65 % of the innovations are considered important in terms of economic development and that just under 70 % of them are regarded as having something to contribute to the solution of ecological problems (see fig. 4.8-1). The assessments for the other areas of the study, however, put the economic importance of these innovations somewhat into perspective, since 65 % constitutes the exact average for the whole study; similarly, the importance ascribed to these innovations in terms of generating work and employment also matches the overall average (20 % in this case).

Figure 4.8-1:   What are the innovations in the field of Energy and Resources important for?

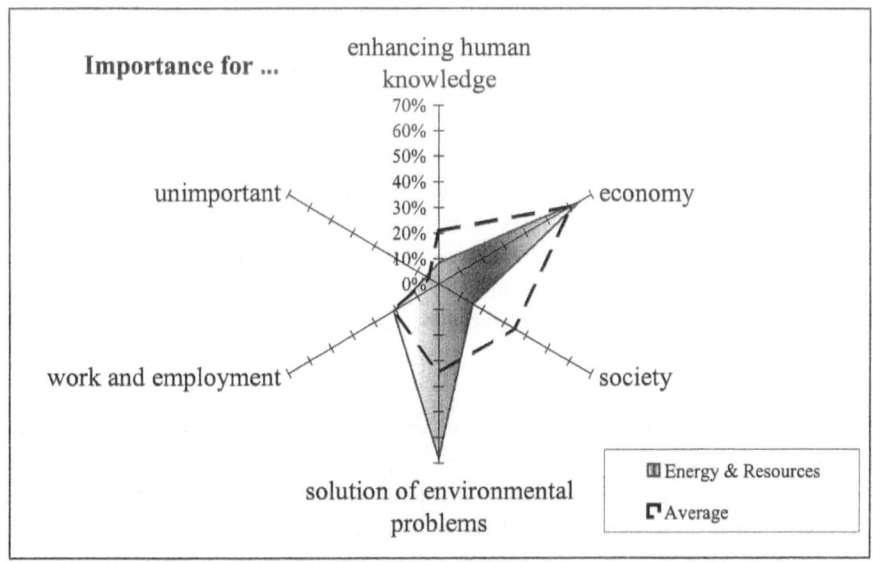

In the subject area of Energy and Resources, twice as many participants expect contributions to the solution of ecological problems than are expected from most of the other sets of topics in this study. This means that new energy technology is not only designed for greater efficiency and substitution but also for environmental protection. The topics mentioned most often in this subject area are *recycling of organic and combustible waste, long-life consumer goods* and *salvage from demolished structures*. The *car consuming only 2 l per 100 km (ca. 1 US gall per 124 miles)* only ranks fifth in this list.

The experts in this subject area see the three main contributions to economic development in the area of raw material production (*prospection methods without test drilling, leaching technology* and *deep-sea mining*).

Since hardly any questions were asked about basic research, the contribution of the innovations in this field to the progress of human knowledge was only cited in about 10 % of the responses, which is half of the average for the whole study. The experts expect the greatest contribution to the progress of knowledge to come from *deep drilling technology* and from new *geological surveying methods*, followed by clarification of the question whether *natural gas generated in the earth's interior can be extracted*. In fourth and fifth places in terms of importance to the advancement of human knowledge are the *operability of nuclear fusion reactors incorporating a solution to the disposal problems* and the *large-scale biological production of hydrogen*. One point that is common to these last three topics is that their realisation would also have a considerable impact on long-term economic and ecological development. Natural gas would become available for an unlimited time, while operational nuclear fusion incorporating a solution to the disposal problem and biologically-produced hydrogen could also help to resolve the issues of exhaustible energy sources and CO_2.

Particular importance in terms of business activity and employment attaches to the topics on the *recycling of commodities* and *salvage from demolished structures* as well as those on the *replacement of chemicals by renewable resources* and the *use of biomass for energy production*. This raises the question whether there was not perhaps an excessive focus here on the activities that are more or less at the experimental stage and still highly labour-intensive, no thought being given to the likelihood that efforts will be made to rationalise these activities before the time of realisation period between 2009 and 2015.

Which of the 114 developments in this subject area are considered unimportant by the expert respondents? The topics most frequently cited in this context were *electricity production from temperature variations in sea water* and from *wave energy*, the *fast breeder reactor* and the *separation of CO_2 from the fumes discharged by power stations*. Nevertheless, all of this technology could - at least in theory - contribute substantially to the solution of the CO_2 problem.

When will these topics become reality?

The median value of the periods, in which the topics in the innovation field of Energy and Resources are expected to reach fulfilment, is the year 2016, which is by far the latest date for any of the fields. The responses also reveal the widest chronological dispersion in the Delphi survey. The main reasons for this late realisation estimate are the average estimates for the topics relating to hydrogen (2022) and for nuclear energy, the production of mineral oil and natural gas, storage mechanisms and fuel cells, for which an average fulfilment date of 2020 is considered realistic. Even the topics from the areas of energy services (2009), substitution by information technology (2012), cross-sectional efficient energy use and recycling (2013) and supply structures as well as buildings (both 2014) are only considered to be medium-term goals. The interquartile range of nine to ten years is fairly constant for all statements.

The practical implementation of methods that would make it possible to *fix CO_2 from fuel-burning power stations* is regarded as a Utopian pipedream by 25 % of the participants and as unimportant by a further 27 %. However, at another point of the same questionnaire, only 13 % of the experts considered it unrealistic to *use separated CO_2 from fossil-fired power stations to increase yields from mineral oil and natural gas fields, incorporating terminal storage*. The average estimate suggests that the separated CO_2 will be usable as early as 2015, whereas the separation technology itself is not expected to be available until 2021. Given current activities in Japan and Norway, are the German experts perhaps underestimating the development of this technology?

What is the level of Germany's research and development in Energy and Resources?

Do German researchers lead the way in this subject area? Or are they just particularly self-important? Or are they perhaps insufficiently familiar with the findings of their colleagues in other countries? Whatever the answers to these questions, more than 60 % of the experts make Germany the top nation (interestingly, Japanese researchers believe the precise opposite, namely that Japan is ahead of Europe). This is the third-highest self-assessment by German experts in the various subject areas, exceeded only by that of the experts in the fields of Environment and Nature and of Mobility and Transport. 45 % of the experts agree that after Germany, the United States is in second place, followed by Japan (30 %) and other EU countries (20 %).

The experts responding to the survey see the principal strengths of German R&D in the areas of buildings, fossil-fired power stations, recycling, supply structures and the efficient use of energy (more than 80 % in each case). German R&D is considered to be at its weakest in the production of mineral oil and natural gas (39 %) and in the extraction of primary raw materials (46 %).

What needs to be done for Energy and Resources of the future?

In more than 40 % of the responses, international co-operation is cited as a particularly effective means of bringing the topics to realisation. The main contributors to this assessment are the high percentages in the areas of primary resources (81 %), nuclear energy (80 %) and the production of mineral oil and natural gas (73 %), areas in which the globalisation of companies is far advanced and in which German R&D tends to score low marks with the experts. The importance of international co-operation is less highly rated, on the other hand, in the areas of buildings (24 %), recycling (32 %) and the efficient use of energy (32 %), in which German R&D enjoys a high standing.

The experts' calls for international co-operation are almost matched by their desire for third-party support. The experts believe that the areas of buildings (58 %), renewable sources of energy (66 %) and fuel cells (66 %) would benefit most from additional funding, whereas only 20 % consider that nuclear energy requires an extra cash injection. As the third most important measure, the experts selected changes in the present regulations, especially in the fields of energy services (82 %), recycling (82 %) and buildings. Scarcely any of the experts, however, thought that changes were needed in the regulations governing primary raw materials, mineral oil and natural gas production or power transmission.

Improvement in the R&D infrastructure and more exchanges of staff between the worlds of industry and science were cited less frequently as desirable aims than in most other fields of the study. The same applies to the improvement of training, which is surprisingly allocated the lowest priority of all the potential means of improving the implementation of the various innovations.

What follow-up problems might ensue?

Although the expert respondents believe that 70 % of the innovations have an important part to play in solving ecological problems, they also consider that more than 35 % of them are likely to bring environmental problems in their wake. Developments in the area of Energy and Resources are thus put on a par with those in the chemicals sector and agriculture.

Figure 4.8-2:   Subsequent environmental problems

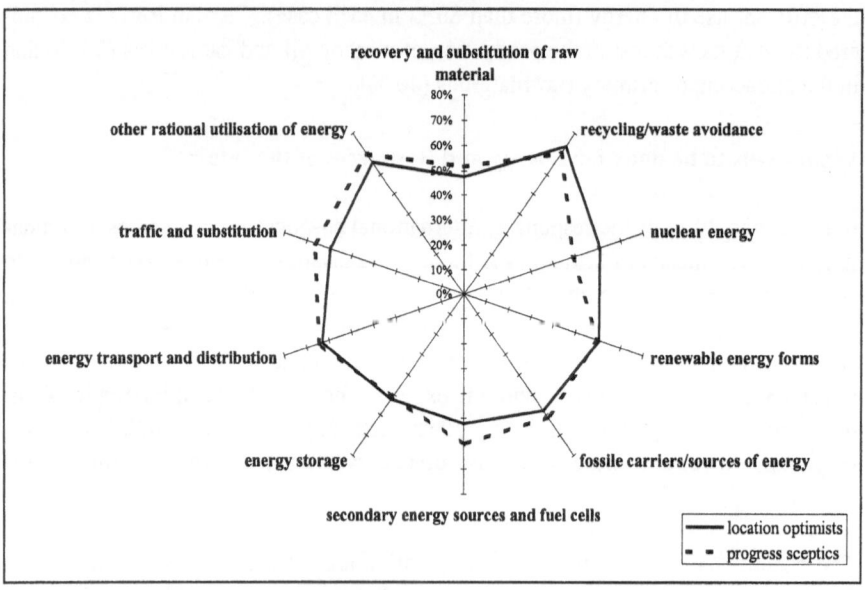

The diagnosis of extreme sensitivity to ecological repercussions among the respondents in this innovation field is verified by some specific examples. 70 % of the experts, for instance, see environmental problems in *scientific deep drilling* and *amorphous solar cells*. This figure equals the percentage of experts who expect ecological problems to result from the *disposal of highly radioactive waste.*

**Key point: Nuclear energy and renewable energy sources will be able to con-
tribute to the reduction of CO_2 levels**

Two of the political hot potatoes among the topics in the subject area of Energy and
Resources are the issues of nuclear energy and renewable sources of energy. Both
of these options are recognised as potential major contributors to the reduction of
CO_2 levels. Some 70 % of the 164 replies express the view that, *because of the
problems of CO_2 emissions and climatic change and because of improved safe-
guards, nuclear energy will come to be accepted throughout Europe*; the experts
with the highest levels of specialist knowledge, as well as the other experts in the
sample, expect this to happen around the year 2015. About five years after that,
according to the vast majority of the expert respondents, *small and medium-sized
nuclear heat and power production plants, which are very safe and suitable for location next to towns and cities, will have been developed and will be commissioned
to serve towns and cities with combined heat and power supplies*. About the same
percentage of experts take the view that *breeder reactors linked into a nuclear fuel
cycle will also have been put into practical use*.

While 90 % of the experts are confident that such uses of nuclear energy are perfectly feasible in principle, serious doubts emerge when it comes to the disposal of
radioactive waste. In one case, indeed, namely the topics that *a disposal method
such as transmutation will be put into practice for highly radioactive solid waste*, a
majority of the experts harbour doubts. Moreover, one-third of the responding experts see no chance of a *globally safe, secure and proliferation-proof system to
confine Plutonium 239* ever becoming *operational*.

The experts believe that *renewable sources of energy (excluding hydroelectricity)
will be responsible for more than 10 % of Germany's power supply* from some time
between 2015 and 2025. This appears possible, since it is estimated that *system
costs for network-linked photovoltaic systems amounting to less than euro
2,045/kWp (DM 4,000) will be achieved* by about the year 2017, that in less than ten
years *wind farms with a capacity of up to several megawatts and specific investment
costs of less than euro 1,025/kW (DM 2,000)* will be *marketable* and hence that in
about 15 years *offshore wind farms with capacities of more than 100 MW will be
built*. And that is not all: in 15 years, *highly efficient energy production from biomass (e.g. gasification of plants, wood and straw) will be widespread*, and even
before then *diesel engines running on cold-pressed vegetable oil will be in use for
many applications in block-type thermal power stations*.

Developments in the nuclear energy sector are most frequently associated with subsequent safety problems, with energy transport and the use of hydrogen and oxygen
following quite some way behind. When it comes to the assessment of social and
socio-cultural repercussions, Energy and Resources is one of the subject areas with
the fewest references to that sort of problem in the entire study. The social conse-

quences of the topics relating to energy services and tariff structures, to substitution by information technology and to buildings technology have the highest ratings in this field. The reason: developments in these areas directly affect people's everyday lives.

It is also interesting that the environmental pessimists, who are likely to be well practised in comparative environmental assessments, anticipate fewer environmental repercussions than the population optimists, except in the areas of nuclear energy and the extraction and substitution of raw materials (fig. 4.8-2).

4.9   Construction and Living

(contributed by Peter Zoche)

Structure of the innovation field

Potential changes in the area of Construction and Living are analysed here with the aid of 75 statements on future innovations. The questions raised by these topics focus on developments in the field of new construction technology and architecture. Possible innovations in the use of materials and construction processes are discussed along with various aspects of building technology, supply and disposal systems as well as the use of information technology. In addition, the topics also refer to questions of redevelopment and resources, and deal with the conditions governing such developments by examining aspects of the quality of life and of safety as well as potential social developments and changes.

If we consider the stages of the innovation process to which the 75 topics should be assigned, just under 18 % of them belong to the innovation phases of clarification and development, almost 35 % of them are devoted to initial commercial applications and 48 % relate to the widespread introduction of the innovations. Some 43 % of the topics in this field of inquiry match those of the Delphi survey that was conducted in Japan at the same time. This is the second-highest agreement rate among the twelve subject areas.

Who was surveyed?

In the first round of this Delphi survey, 110 people completed the questionnaire for this field of inquiry; the figure for the second round was 94. About two-fifths of the participants work in businesses. The group of respondents who work in universities, the civil service or private non-profitmaking institutions is equally large, while about one-fifth of the respondents are employed elsewhere.

The respondents working in universities and in the civil service rate their own specialist knowledge considerably more highly than company employees. It should be emphasised that, compared with the other fields, the realm of Construction and Living boasts a disproportionately large number of respondents who credit themselves with a high level of specialist knowledge.

The experts consider themselves well qualified to assess statements relating to changing architectural and settlement structures, such as the statement that: *newly developed models of land taxation with higher charges for the utilisation of larger surface areas lead to compact settlement structures and hence the avoidance of vehicular traffic*, that *variable methods of housing construction permit individualised housing design, even in multi-storey buildings* and that *car-free residential areas and populated areas best suited to communal vehicle use are widespread*. These are topics that also feature prominently in the public debate.

The topics on which particularly large numbers of experts claim to have little specialist knowledge relate to *space stations*, *extinguishing and rescue technology for fires in high-rise buildings* and *systems to guide and orientate the visually impaired on pavements and footpaths, using sensors and synthetic speech*. However, the replies from the respondents with less specialist knowledge were substantively similar to those from the experts with a high level of specialist knowledge in this area.

Why is Construction and Living an important field?

An unmistakable trend emerges from the experts' assessment of the importance of innovations in the subject area of Construction and Living (fig. 4.9-1), economic development being mentioned most often, followed by the solving of ecological problems and social development.

The experts employed in industry made a more optimistic assessment of the impact of the innovations on economic and social development and on the preservation or expansion of business activity and employment than the experts from the world of science. Innovations in the field of Construction and Living are about average among the subject areas in terms of their perceived importance in generating business activity and employment, but they are thought to have scarcely any significant effect on the advancement of human knowledge.

Figure 4.9-1:   What are the innovations in the field of Construction and Living important for?

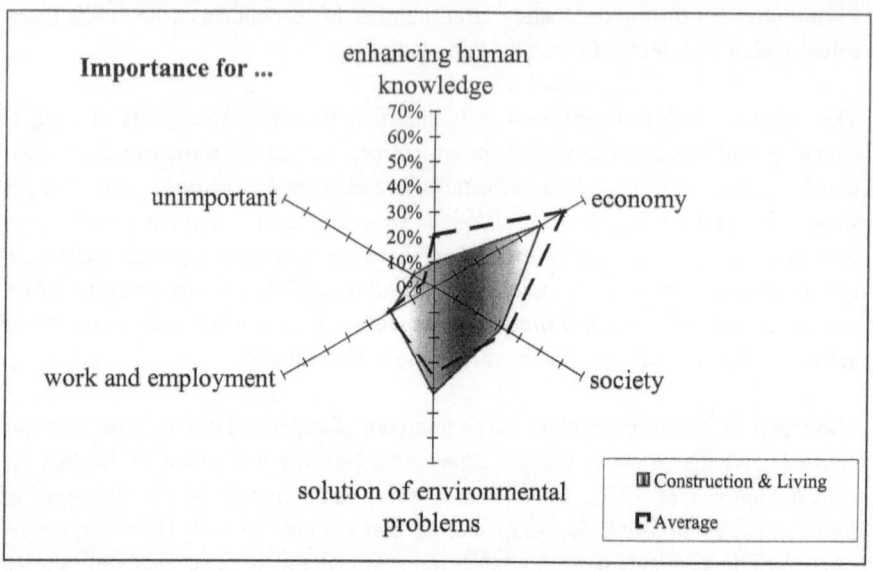

When will the topics become reality?

In the experts' opinion, around 13 % of the innovations will be possible in the period from 2001 to 2005, roughly 48 % in the years from 2006 to 2010 and some 29 % between 2011 and 2015. This means that the topics in the area of Construction and Living can be expected to become reality at a slightly earlier time than the average estimated time of realisation for all innovations in the twelve subject areas under examination.

The innovations that are presumed to be closest to implementation are that *the technical conditions for facility management systems will be incorporated into the construction of buildings* as well as the use of *control systems and the associated sensor technology which keep the excavated cross-section exactly the same for the tunnel cross-section and therefore ensure that losses are minimal.* The experts' average assessment for each of these suggests that they will be implemented in the year 2004, and the interquartile range is relatively narrow, at two years below the median to three years above. So the experts' assessments are largely in agreement on this point. The three most improbable innovations are considered to be *the clarification of the effect of ultra-tall buildings on people, concepts for assessing mental sensitivity and physiological responses to the built environment and the unmanned building site.* The experts also see relatively slim prospects of fulfilment for the *use of special construction technology for skyscrapers (about 1,000 metres high) in*

Germany and for the proposition that *building materials used will consist mainly of new synthetic materials which are 100 % recyclable.* These topics are considered unrealistic by 25 % and 19.4 % of the experts respectively.

What is the level of Germany's research and development in Construction and Living?

According to the German experts in the survey, research and development in the subject area of Construction and Living is furthest advanced in Germany, followed by the United States of America and Japan (fig. 4.9-2). No other countries inside or outside the European Union were considered to be playing a leading role in relation to any of the topics. In this subject area too, there is an interestingly high level of concurrence between the views of the experts with a high level of specialist knowledge and those of the whole sample of experts. The leading position ascribed to Germany by the community of German specialists, however, is not corroborated by the Japanese Delphi experts, who see Japan at the forefront of R&D with the United States in second place, while the (Western) European countries come a poor third in the international rankings. The leading role of Japanese companies in the realm of automated construction undoubtedly has something to do with this assessment.

Figure 4.9-2:   The international research and development situation

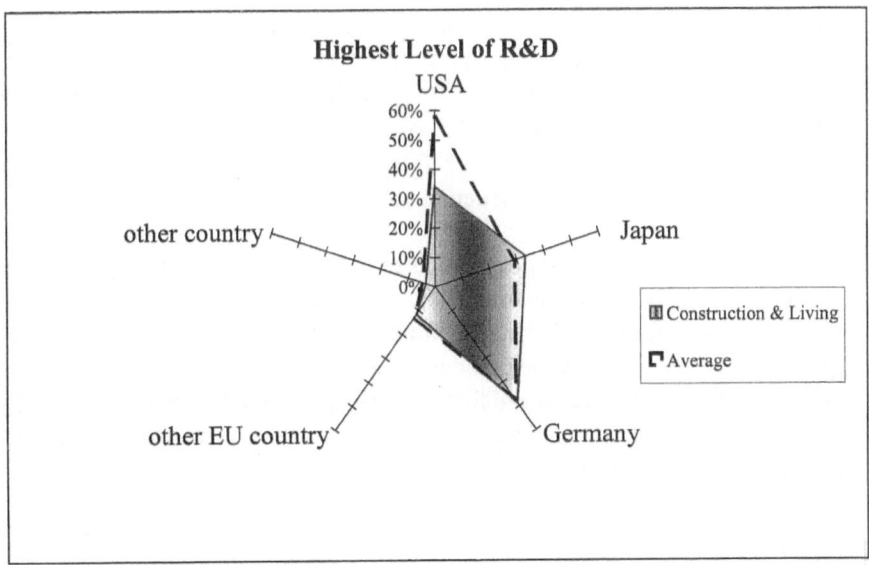

In the judgement of the German experts, the more detailed picture of the international research and development situation looks like this: American researchers lead the way in the realm of quality of life and safety (*a significant percentage of the*

working population will take advantage of a range of residential services which extends to "hotel-type" service) and in the area of building technology *(the technical conditions for facility management systems will be incorporated into the construction of buildings)*. In the area of new building techniques and architecture, Japan plays a leading role, e.g. in efforts to ensure that *progress in water engineering technology will make floating cities at sea possible* and that *buildings will be erectable quickly and cheaply using a fully-automated process (unmanned building sites)*. These are topics which probably imply a considerable research requirement in the field of information technology.

The experts give Germany a clear technological lead in the spheres of resources and redevelopment *(self-financing recycling technology will be developed in the building industry)*. *Environmentally-friendly forms of technology, which avert the destructive influence of pollutants, will be introduced for the conservation of historic properties. Technology will be introduced in Germany which permits 100 % recycling of by-products from construction work, such as concrete, asphalt and excavated soil)*. In the field of supply and disposal systems *(progress in efficient energy use resulting from long-term power or heat storage and the use of renewable sources of energy will make it commonplace for newly-constructed industrial, commercial and residential buildings to supply their own energy. The use of final forms of energy and the recycling of waste will be widespread in Germany's local communities)*. And in the development of materials and construction processes *(use of mainly recyclable plastics in building materials or tunnel-building machinery, with which the rubble from the excavation process is made immediately reusable)*. The various topics on the development of materials and construction processes are not so dissimilar to those of our competitors in Japan and the U.S. as is the case with the other areas of Construction and Living that were discussed above.

What needs to be done for Construction and Living ?

When asked for the measures they considered to be particularly important in terms of helping to implement the innovations to which the 75 topics relate, the experts put improvement of the German research infrastructure in first place, followed by regulatory changes and improved training. The only innovation field in which the experts identify a more urgent need for training is that of production. Lower priority is attached to international co-operation than in any of the other subject areas covered by the Delphi survey (cf. fig. 4.9-3).

Figure 4.9-3:   What measures should be taken?

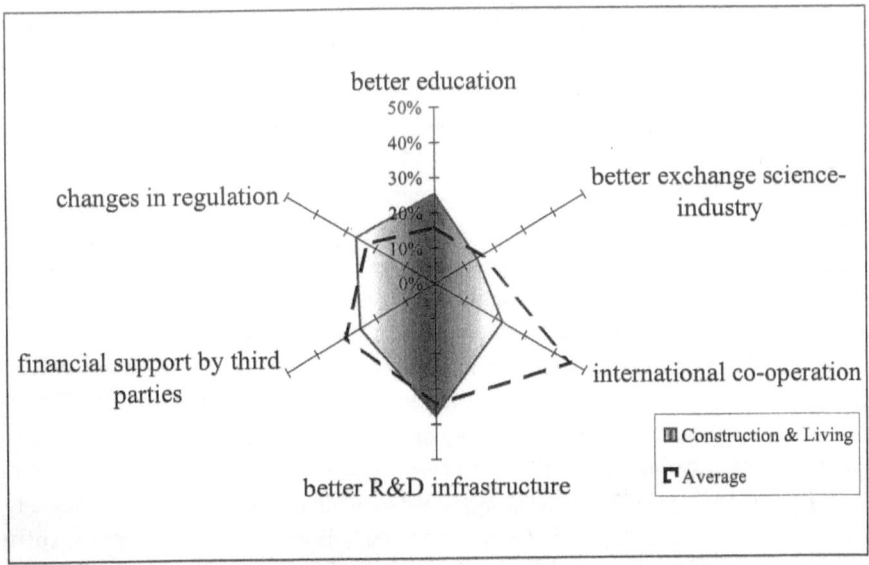

The average ranking of staff exchanges between science and industry in order of importance assumes particular significance insofar as it is often cited in connection with the improvement of the training situation, for example in the following four topics:

- *The entire construction process will be administered by means of a networked (tele-) information association involving all participants.*
- *Site meetings will be conducted with the aid of teleco-operation systems.*
- *Building processes will be standardised, and the organisation of site management through control of even the smallest segment of the building process will be optimised.*
- *The technical conditions for facility management systems will be incorporated into the construction of buildings.*

What follow-up problems might ensue?

The experts cite environmental consequences as the main problem that could ensue from the implementation of the innovations in this subject area. Such consequences, however, are mentioned by only about 40 % of the sample, which is roughly average if we compare this subject area with the other eleven in the Delphi study. The fields in which innovations are generally considered most likely to create subsequent problems, are those of *construction technology and architecture*, and *new materials and construction processes*.

> **Key point: Major progress soon, but Utopian visions meet with scepticism**
>
> In the area of Construction and Living, new visionary ideas and strategies relating to the residential environment of the future are greeted with the least enthusiasm imaginable. Utopian plans, such as the construction of ultra-tall skyscrapers or of *space stations, in which even untrained people live for more than a year*, meet with the approval of scarcely any of the expert respondents. Not even the automation of the construction process arouses much euphoria in Germany - here Japan and the United States will surely preserve and increase their lead.
>
> On the other hand, the German experts highlight a wide range of innovations which can be implemented quite rapidly and which can make valuable and creative contributions to future economic and ecological development in the area of building. Examples of such innovations are the extensive use of *facility management systems* and the wealth of development opportunities for new applications in the field of resources and redevelopment, in supply and disposal systems and in the development of materials and construction processes. These priority areas offer scope for socially and ecologically relevant applications which, if marketed internationally, could be of great benefit to the German economy, especially as the German experts consider that their country's R&D enjoys an international advantage in terms of know-how.

Possible repercussions in the social, cultural areas or other areas of society tend to merit few mentions in the context of Construction and Living compared with other fields of innovation. Nevertheless, some specific topics do attract noticeably adverse assessments. Among these are the spread of *facilities in the homes of ordinary German families whereby the experiences of travel, film performances, sporting contests etc. can be enjoyed using virtual reality* (98.4 % expect this to have social, cultural or other repercussions) and the *disappearance of the distinction between work and the home through the emergence of "virtual companies"* (98.3 %).

4.10   Mobility and Transport

(contributed by Peter Zoche)

Structure of the innovation field

Prospects in the research field of Mobility and Transport are described in 107 statements. They are clustered around six different groups of issues and range from vehicle engineering through safety and environmental factors, various transport systems and the use of computerised communication systems ("telematics") in private and public transport to general questions of mobility and consideration of determinants of traffic volume.

With regard to the stages of the innovation process to which the 107 topics can be assigned, about 17 % of them belong to the clarification and development phases, about 21 % are devoted to initial commercial applications and 62 % of the topics - the second-highest percentage in any of the subject areas under examination - deal with the widespread dissemination of the innovations. Only 17 % of the topics in this field of inquiry match those of the sixth Japanese Delphi survey. This is the lowest agreement quotient within the twelve subject areas of the German Delphi survey.

Who was surveyed?

In the first round of the Delphi survey, 150 persons completed the questionnaire on Mobility and Transport; 122 of them responded in the second round. A total of 48 % of the participants are employed by companies - more than in any other subject area. 34 % of the participants work in universities and a further 9 % in the civil service or in private non-profitmaking institutions.

For 85 % of the topics, more than half of all respondents rated their specialist knowledge as low. For about one-third of the topics, the number of experts with a high level of specialist knowledge exceeds 15 % of the total sample. The highest number of "specialists" (23.5 %) emerged in connection with the statement that *telecommunications systems will be put into widespread use to achieve an intelligent distribution of traffic and transport among the various traffic routes and transport systems, to use the existing transport infrastructure more effectively and more efficiently, to assist in the elimination of bottlenecks and peak periods and to encourage or permit the use of flexible means of transport.*

The topics about which the respondents tend to have a high level of specialist knowledge also elicit an above-average number of responses; with 116 responses, the statement on *motor vehicles with 30 % less petrol consumption* attracted the largest response quota. Conversely, where there were fewer responses, there were also few experts with a high level of specialist knowledge. For instance, the question on the *creation of space stations with a corresponding requirement for cost-effective new propulsion and transmission architecture to meet the special requirements of the space environment* had the lowest number of specialist respondents as well as the second-lowest number of responses.

Particularly few responses were received to the question concerning improved *ship-building materials and ship engines.* Most of the experts who responded to it are from other sectors of industry. Nonetheless, the responses of the "top specialists" to this topic, as they do on average across all subject areas, differ only insignificantly from those of the other experts.

Why is Mobility and Transport an important field?

The main contribution of the innovations in the field of Mobility and Transport, according to the experts questioned, will lie in solving ecological problems (approximately 55 % of the replies); there are only two fields in which the experts exceed that level of estimates with respect to the solution of ecological problems. A total of 53 % of the experts highlight the importance of the topics in relation to economic development (fig. 4.10-1).

In relation to the other innovation fields under examination, this figure was close to the average. The contribution to social development received an above-average assessment, being endorsed by some 33 % of the respondents. On the other hand, the selected topics in this subject area are not perceived to contribute much to the creation or preservation of jobs or to the progress of human knowledge.

Figure 4.10-1:   What are the innovations in the field of Mobility and Transport important for?

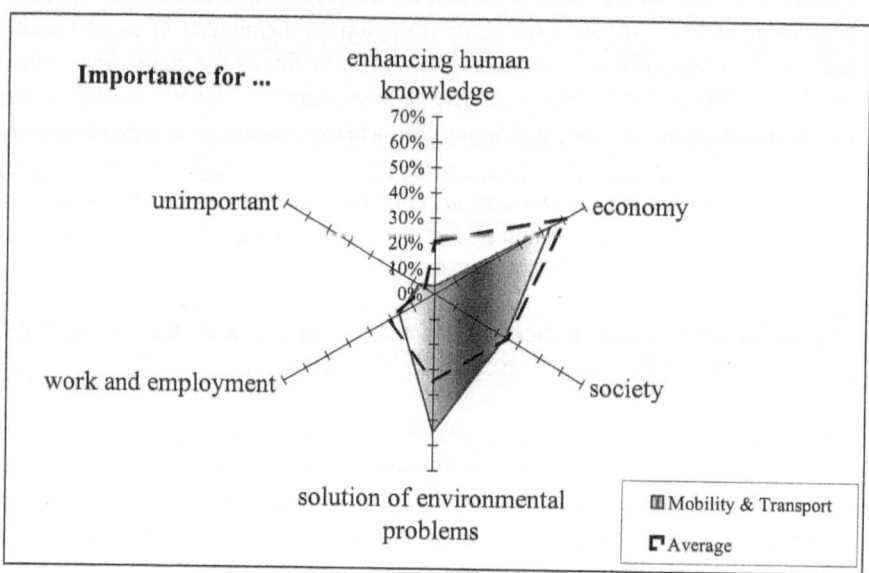

When will the topics become reality?

In the estimation of the participating experts, the innovations in the field of Mobility and Transport will take less time to implement on average than the median time of realisation for all topics contained in the Delphi survey. Some 70 % of the topics might become reality by the year 2010, but most of them are expected to do so after 2005. A further 20 % of the innovations shown could be implemented in the years from 2011 to 2015, and only 10 % are assigned to the period between 2016 and 2025.

Logistical, environmental and consumption-optimisation modelling are expected to clarify between 2001 and 2008 whether *the most efficient speed limits for passenger vehicles are 120 kmph on motorways, 80 kmph on country roads and 30 kmph in built-up areas.* This topic could be the first to reach fulfilment, although the comments on it are quite conflicting, with more than 20 % of the respondents taking the view that the innovation outlined will never be implemented. By contrast, the experts concur in their assessment of the topic on the widespread proliferation of *electronic banking in households* and on the *possibility of selecting and purchasing many goods by PC, thanks to virtual reality and multimedia (delivery or collection not tied to opening hours).* These, in the opinion of the experts, can become reality within a narrower time-frame (2002 to 2004 and 2002 to 2008 respectively), and hardly any doubts are expressed about their feasibility.

In the experts' estimation, the following are the most improbable innovations, 61 %, 53.9 % and 49.3 % respectively believe that the innovations can never be implemented:

- *On the basis of new economic findings, doing without a car will be rewarded through special bonus systems, e.g. in the form of discounts on private electricity bills.*
- *Every car driver will receive an annual mobility account card for journeys into or in town; if passengers are carried, the driver will receive an ecobonus in the form of a tax refund.*
- *Individually-owned light aircraft will account for a considerable proportion (up to 5 %) of passenger travel over medium distances.*

What is the level of Germany's research and development?

In the innovation field of Mobility and Transport, Germany plays a leading role in the world across the whole spectrum of subject groups in the opinion of the German Delphi experts (61 %, cf. fig. 4.10-2). The United States (31 %), Japan (28 %) and the other countries of Europe (19 %) occupy second, third and fourth places. Only in two other fields (Environment and Nature, Energy and Resources) is Germany judged to be even stronger.

Figure 4.10-2:   The international research and development situation

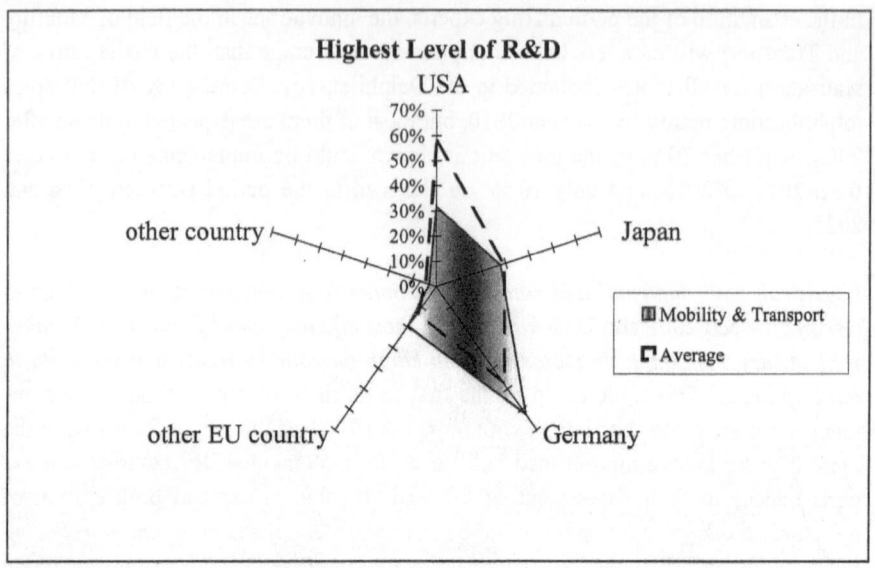

What needs to be done for Mobility and Transport of the future?

The measures called for by the expert respondents to foster innovative potential are of a decidedly regulatory nature. Only in the field of Environment and Nature is there a comparable number of calls for regulatory measures. Another characteristic of the Mobility and Transport field is that all the other types of action put forward for discussion are less frequently advocated by the experts than in any of the other eleven subject areas (fig. 4.10-3).

Regulatory measures are considered especially important in relation to the following topics (the figures in brackets refer to the percentage of experts who believe that regulatory changes are required):

The indirect beneficiaries of improved access to local public transport will contribute to its running costs through a combined agglomeration and deglomeration charge. The agglomeration charge will skim off the advantage enjoyed by those who live in densely-populated areas in terms of blanket transport coverage and frequency of services. The deglomeration charge will skim off the higher running costs that result from dispersed settlement structures. (100 %)

Every driver will receive an annual mobility account card for journeys into or in town; if passengers are carried, the driver will receive an ecobonus in the form of a tax refund. (96.8 %)

The funding of local public transport will be completely reorganised. The infra-structure costs will be met from a local transport charge and from a local transport development charge, and the current development payments for roads and car parks prescribed in the Code of Building Law (Baugesetzbuch) and the law governing local rate levies will be reduced or cease to apply. (95.9 %).

Figure 4.10-3: What measures should be taken?

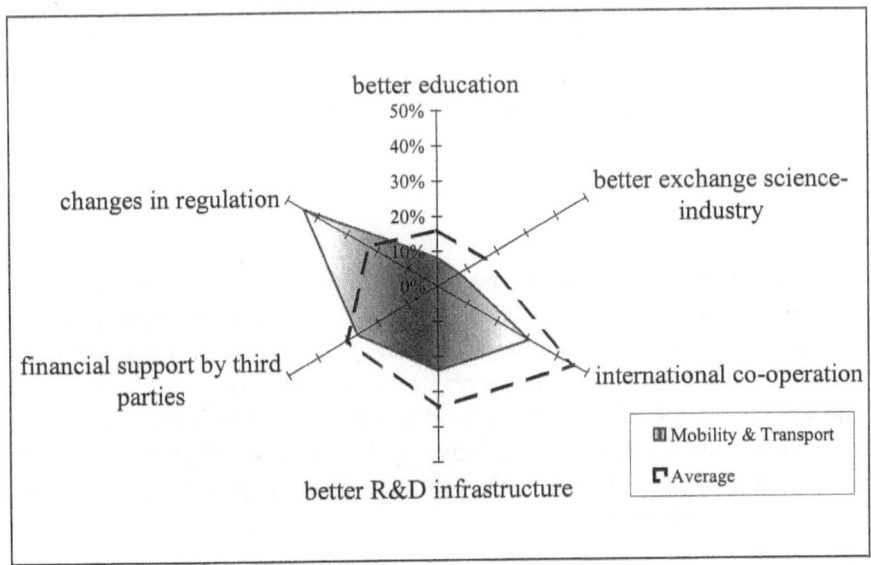

International co-operation is considered important for topics such as the *compatibility of national high-speed rail networks (e.g. TGV and ICE)* (90 %) or *hypersonic 300-seat aircraft with speeds of Mach 2.5 on international scheduled flights* (86.5 %). Third-party support is cited most frequently in connection with two topics, namely *local public transport will be made fast and attractive by the use of small vehicles operating on demand at off-peak times* (75.8 %) and *customer-friendly changeover points will make transfers from private to public transport and between buses and trains an attractive option; the area around railway stations will be redesigned as a world of appealing sights and sounds* (66.3 %). Compared with the average for the twelve innovation fields, improvement of the research infrastructure is considered less important. The experts attach even less importance to improved training and to more exchanges of personnel between industry and science (see fig. 4.10-3).

What follow-up problems might ensue?

Fewer potential problems are foreseen in the implementation of the innovations referred to above than is the case with other areas of technology. Nevertheless, two points should be emphasised: firstly, the topics relating to vehicle technology and transport systems are considered by the experts in the survey to pose most problems; secondly, safety issues are raised more frequently in the field of Mobility and Transport than in most other areas; safety problems are anticipated, for instance, in connection with *a system which, with the aid of laser beams or ultrasound waves, discovers obstructions such as persons or vehicles on railway lines and causes approaching trains to stop automatically* and *driver-support systems which receive essential information and signal it to the driver or modify vehicle functions.*

Key point: Even though mobility increases, the environment can still be protected

The ever-increasing mobility that our society demands is highly ambivalent. It is associated with economic and personal advantages but is equally regarded as personally threatening and ecologically destructive. The statements formulated in this field hover in the magnetic field between these two poles. Nevertheless, individual innovations can result in the protection of resources and in the creation of an ecologically desirable situation in the transport sector. Moreover, the potential ecological contribution of innovations in the field of mobility and transport is greater than in any of the other subject areas surveyed.

The experts' undivided endorsement is given in particular to those environmentally-friendly innovations that are primarily based on technological advances, such as the creation of engines with lower fuel consumption or improved exhaust technology. Measures relating to the transport system, such as tipping the balance in favour of public transport by means of traffic management in mobility centres or through the increased use of computerised communication systems in transport operations, elicit a more ambivalent verdict. The experts do, however, generally regard such system-related solutions as particularly important in terms of economic development. For all that, the ability of many of the proposed solutions to mobility and transport problems to gain acceptance will depend entirely on the adaptation of the regulatory framework. And the respondents are largely sceptical about the extent to which the proposed ideas will prove achievable under present circumstances. In other words, political decisions need to be taken about the future.

4.11   Space

(contributed by Ulrich Schmoch)

Structure of the innovation field

Since the initial boom phase of the 50s and 60s, space travel has undergone many ups and downs to emerge as an established field of technology. As far as its frame of reference is concerned, the distinction must be drawn in space travel between satellite technology (including the application areas of earth observation and tele-communications) and space laboratories close to the earth for the purposes of re-search, production and the exploitation of resources. Other branches are deep space missions as well as transport systems. There are partial overlaps between the four branches mentioned.

After several decades in which the emphasis in space travel has been placed on re-search, what is now important is the concrete implementation of the discoveries made. Out of a total of 78 topics in the survey assigned to space, there are (still) only very few which refer to the fundamental elucidation of principles or phenom-ena. With a share of more than 50 %, the statements on practical application occupy a very broad spectrum.

Who was surveyed?

The number of experts (78) who took part in both rounds of the survey was the low-est in comparison with the remaining innovation fields of the Delphi survey: the "community" of space (travel) researchers is limited. With respect to the institu-tional origin of the experts, the distribution however corresponds roughly to the average ratios of the study.

Why is Space an important field?

Despite the tendency towards application, in the estimation of the experts ques-tioned "space" focuses its attention on the twin goals of the expansion of human knowledge and economic development. Although only a few questions concern the acquisition of scientific knowledge, their significance is well above the average figure for all innovation fields. As far as economic significance is concerned, roughly the average level is attained. Of the aspects of "social development", "work and employment" and "solving ecological problems", at best a greater significance is attached to the latter (fig. 4.11-1). Remarkably, the experts from industry and science essentially agree where these basic assessments are concerned.

Figure 4.11-1:   What are the innovations in the field of Space important for?

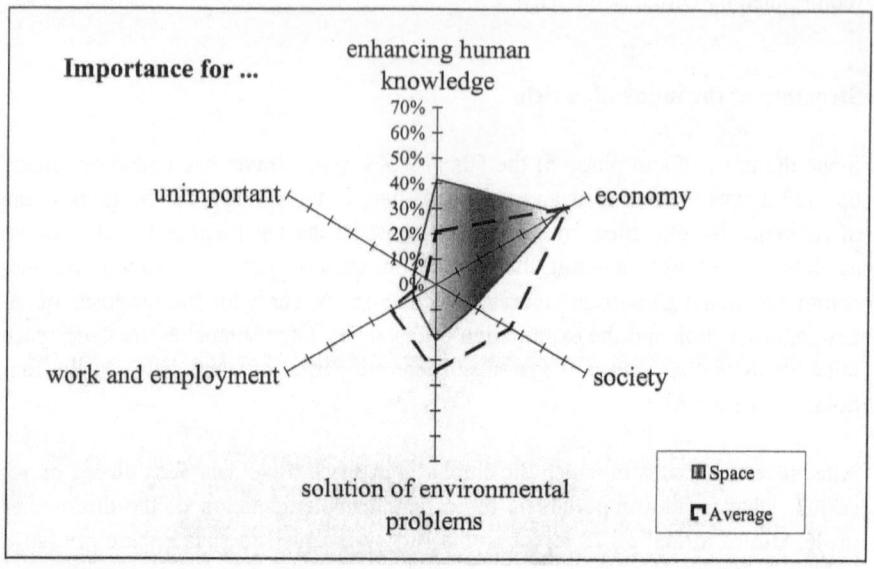

An analysis by subject groups or branches of space travel shows that an expansion of human knowledge can be expected above all from deep space missions, for instance from the *landing of a manned spaceship on Mars* or an *investigation of Mercury with the aid of satellites*. At the same time, the experts class the economic significance as extremely low, precisely where these topics are concerned. On the other hand, the aspect of economic significance is stressed in satellite technology and its application for navigation and telecommunications and in space shuttles operating close to the earth. Where these topics are concerned - in a mirror image to the deep space missions - negligible significance for the progress of human knowledge is evident.

The economic significance of production and resource exploitation in space is classed by the experts as only average overall, whereas a comparatively high level of importance is attached to the *automatic production of alloys and crystals in near-earth orbit* as well as to *solar power stations in space*.

The significance of space travel for solving ecological problems is certainly rather small in comparison with other fields. The experts however expect a substantial contribution to environmental protection from earth surveillance by satellites.

Taken as a whole, the various branches of research in and about outer space can be characterised very well via the importance criteria. However, only limited comment can be made on the latent controversy as to whether manned or unmanned space travel should be favoured, because alongside questions of economy, above all aspects of reliability, safety and the long-term nature of missions are involved. That this question has a very high status with the experts is shown by the fact that the topic concerning the *complete replacement of manned by unmanned space travel* provoked the greatest number of comments. Overall, agreement emerges to the effect that in the foreseeable future there will not be a complete replacement of manned space travel, but probably a clear reduction in favour of unmanned space travel. To the supplementary statement concerning the *availability of automatic systems for conducting experiments and maintaining orbital infrastructures* the experts at any rate attach very high economic significance.

When will the topics become reality?

When compared with all innovation fields, the time for the implementation of the space travel topics is very late on average: in second last place at 2013. This can be put down to the fact that (as in all fields) no fulfilments are assumed for any of the topics before the year 2000, whereas for a comparatively large number implementation is assumed after 2020 or even 2025. The already-mentioned focus of the statements on the application of systems and procedures in practice thus does not mean that their realisation is directly imminent. Rather, the experts base their assumptions on a substantial research and development requirement before practical applications can take place. At the same time, the main emphasis in the coming years will be on applied research.

With regard to the time horizon for the realisation of the topics, distinctions must again be made by groups of topics or branches of space travel to obtain an adequate picture. In the case of satellite technology for earth observation and telecommunications which - as discussed above - is interesting both from a commercial and an ecological point of view, the Delphi participants expect relatively early implementation: on average in the period between 2005 and 2011. On the other hand, deep space missions, which are of interest principally for the expansion of human knowledge, are expected to come to fruition very late. The medians lie in the period between 2015 and 2025, some even later. Herein lies the decisive reason for the - on average - relatively distant time horizon of space travel. If they were limited to satellite technology and its applications, the average figures of the remaining subject areas would be reached.

Even for production or resource and energy exploitation in space, to which at least a medium level of economic significance is ascribed, the assumed realisation time is on the whole very late. Whereas the *automatic production of alloys and crystals in near-earth orbit* is at least considered realistic in the year 2015, all other future projects are dated on average to the year 2025 or later. The latter applies for example to *solar power stations in space* as well as to *production and resource extraction on the moon*.

In view of the multitude of visions which are dealt with in the media concerning the future use of the outer space, in the final analysis it is also instructive to note which topics the experts regard as Utopian. For none of the statements in the field Space is there a majority of experts who regard it as completely unrealistic. The sole exception is the already-mentioned *complete replacement of manned by unmanned space travel*, which at least 60 % of those surveyed regard as unrealistic. On top of this, there are some topics which at least 20 to 30 % of the experts assume will never come to fruition, and for which the remaining experts estimate a very late time of realisation. Among these are *the establishment of space tourism as an economic factor* or *space funerals*.

However, even less curious things such as *production and resource extraction on the moon* are regarded as Utopian by a substantial number of experts. It emerges from the relevant comments that it is less technical than financial problems which are considered to be the limiting factor. Finally, a large number of those surveyed see no prospects of fulfilment for the *dumping of radioactive waste and other extremely toxic waste in space*, placing safety considerations and questions of cost to the fore. The large number of comments on these topics can be judged as an indicator of the fact that this issue is currently the subject of controversial discussion in expert circles. In addition, large *electromagnetic accelerators for transporting materials in orbit* and *microfusion drives for spacecraft* are consigned by many of the experts questioned to the realm of Utopian fantasy.

What is the level of Germany's Space research and development?

On account of the considerable resources invested by the United States in space travel over the last few decades the USA, in the estimation of the experts surveyed, understandably hold a clear lead in research and development. However, when it comes to individual statements, some experts also see Germany as having a leading role. At just under 20 %, this proportion ranks as the lowest compared with all the subject areas of the Delphi survey. A leading position is ascribed to Germany particularly often when it comes to topics on the application of satellites in earth surveillance and telecommunications - branches which are of particular commercial interest.

What needs to be done for Space research?

Alongside the expansion of the R&D infrastructure and financial support, strengthened international co-operation clearly ranks first among the measures demanded for the future. One fundamental reason for this is - as might be expected - the enormous costs of space travel systems. Even now 75 % of Germany's funds go to the European Space Agency (ESA) and, over and above this, collaboration takes place with international partners on 80 % of national activities. Moreover, many programmes relating to earth observation are directed at global systems whose implementation is only meaningful on an international scale. The pivotal significance of international collaboration programmes is also evident from the fact that it is mentioned more often in connection with the space travel topics than in relation to all other fields of the study.

Against the background of a German research position which, by and large, is weak, there is a reminder first and foremost of the need for an improvement in the R&D infrastructure and greater financial support in order to bring about a change in the existing bottlenecks affecting German support funds for space travel which have been falling in real terms since 1993. The call for more public funds for travelling into outer space, which is heard more loudly than in the other innovation fields, refers here not only to deep space missions and thus the expansion of human knowledge, but also to the various areas of earth and environmental surveillance.

At the same time, the experts make the assumption that already by the year 2005 *applications of satellite technology which are not primarily scientific* will be *marketed generally on a private enterprise basis,* and they attach a very high significance for economic development to this point. Against this background, it is an important task of research policy to define the division of labour between public and private participants for the future and, particularly as far as the application of satellite technology is concerned, to make concrete plans for and to negotiate the transition from state to private responsibility.

What follow-up problems might ensue?

In comparison with the other innovation fields, hardly any potential problems are mentioned in the context of the field Space. However, attention is drawn to the fact that greater environmental problems will occur, should resource and energy exploitation in space assume greater proportions. According to the comments of several experts, the environmental damage caused by increased space shipments would outweigh the ecological advantages gained from resource savings on earth.

Apart from this, problems are only mentioned in the context of individual topics. Striking examples are possible environmental problems caused by *supersonic rockets/ aircraft for earth-earth and earth-orbit connections* and social consequences of a substantial increase in efficiency of telecommunications using *satellites,* for instance *with broadband, long-distance or digital transmission.* In the main, safety problems are seen in the use of *nuclear reactors for supplying energy for drive purposes* and in the use of *space as a waste disposal site.*

> **Key point: Inner space travel will become increasingly important for the economy and the environment**
>
> Under the umbrella of Space a catalogue of Delphi topics is processed relating to research and use. The recognition that research and use have to be judged very differently runs as a central thread through the many individual assessments. Deep space travel assists in the expansion of human knowledge and connects with an age-old dream of mankind to learn more about the cosmos. The progress of knowledge as a cultural asset drives this development forward. The landing of men on Mars or the investigation of Mercury have no economic significance.
>
> On the other hand, inner space travel (close to the earth) will become increasingly important for the economy and the environment. Satellite technology for earth observation and telecommunications is attractive both from a commercial and an ecological viewpoint. Considerable progress is expected in approximately 10 years. As far as this commercially important aspect of space research is concerned, Germany is today recognised as having a good level of performance, whereas in other respects the United States are way ahead. Nevertheless, it must not be overlooked that many aspects especially of earth observation are aimed at global systems which can only be implemented meaningfully in an international context. The experts assume that commercial applications of satellite technology will generally be run on a private enterprise basis in just around 10 years. Therefore, research policy has the important task of negotiating the division of labour between public and private participants, and of defining it with an eye to the future.

4.12   Big Science Experiments

(contributed by Hariolf Grupp)

Structure of the innovation field

The subject area of Big Science Experiments deals with an assortment of major equipment, experiments and their applications. It stands out from the other fields of inquiry of the Delphi survey and by the nature of the subject is dominated overwhelmingly by pure research questions. The interfaces to future economic practice only figure to a minor extent. This is within the entire purpose of the survey and makes it possible, more precisely than at other points in this report, to take note of differences in the response behaviour of basic researchers and business practitioners.

The range of the field, covering only 50 topics, may appear narrower than other fields of this survey. Viewed from the internal perspective of scientific systematology however, the field appears very large, encompassing as it does several disciplines. It involves astronomy, but also experiments on our earth (geo-experiments), it covers both elementary particles and nuclear structures as a goal of knowledge, together with the utilisation of major experimental equipment (neutron and synchrotron sources, intense-field laboratories, high-performance lasers and accelerators) as aids for "small-scale" research in solid-state physics, the material sciences, chemistry, biology or medicine. A politically controversial field is included in the form of nuclear fusion. The fact that major experiments of all kinds cannot be conducted without sophisticated measurement and verification technology forms a further topic of the survey. The evaluation of such experiments requires large and ultralarge computers together with computer simulations. This is why the research field of "computational science" is also included.

Within the scientific communities a restrained but critical dialogue concerning the purpose and status of major research can definitely be noted - as it can among the experts who helped to prepare the catalogue of Delphi topics. If there were unlimited research funds, probably no-one would call into question major equipment research. However, as a few "large" sums are in competition with a lot of "small ones", science questions the priorities and along with them the "retrospective priorities". In the field, an attempt was made to create a balanced ratio of the somewhat conflicting specialist interests, by including alongside blocks of subjects on the major experiments for their own sake (knowledge-oriented) a clear emphasis on the use of major equipment as aids for other research.

From a cross-comparison of the innovation fields it can be ascertained that at almost 40 % the proportion of pure research questions is higher than in any of the other subject fields. Approximately the same number of topics is devoted to technical development, i.e. for example to the creation of prototypes. Only somewhat more than 20 % of the statements deal with the application of technologies - in fact predominantly first applications, not widespread dissemination. The field thus emphasises the main points of science and technology.

Who was surveyed?

Despite the orientation towards pure research, more than 40 % of the responses in this field were received from private sector companies. The public sector contributed 55 % of the responses. Most of these experts come from universities or from big science research institutes. The proportion of experts from industry in the field of Big Science Experiments is not lower than in other fields. How should this be interpreted? Single items of laboratory equipment for major equipment research are manufactured by interested industrial firms from time to time. They are not only used for conducting sophisticated experiments, but raise the level of technical knowledge of the manufacturing company above that which can actually be sold on the market. Scientific researchers speak of "research technology" and the "equipment connection of theories". Pure research therefore does not only influence applied R&D; some industries definitely have a practical interest in the progress of big science experiments at least via the instrumentation of big science equipment.

The avid interest of industry in big science equipment research is significant and commendable. However, the representatives of industry have a distinctly lower level of subject knowledge in this field than the scientists. The experts with the highest level of specialist knowledge come from universities and private non-profitmaking research institutes (essentially big science research centres of the Helmholtz group). A relatively high level of technical knowledge of the experts from industry is found in the area of instrumentation - as the above interpretation confirms.

Why is Big Science Experiments an important field?

Against the structural background of the subject area it is not surprising that the assessment of economic significance turns out to be lower than in other specialist fields of the Delphi survey. However, the importance of big science equipment research for social development or for work and employment is assessed as even lower than the economic relevance. What is paramount is the expected contribution to the expansion of our knowledge, i.e. to the advance in knowledge. The experts attach such a high degree of importance to no other field in this respect. The sharply-defined profile in the various dimensions of importance is outlined in figure 4.12-1.

The Delphi participants expect the greatest advance in knowledge to come from the solving of the *problem of dark matter in the universe* - one of the most important issues in modern astronomy. Subjects from the field of astronomy also occupy the next two places: the *experimental verification of gravitational waves* and *the clarification of the cause of the violation of the symmetry principles*. The solution to the *problem of dark matter in the universe* could be found in the second decade of the next century, along with the *experimental verification of gravitational waves*. However, the spreads of the assumed realisation times are large in both cases.

On average, the topics of the block of subjects entitled Big Science Experiments are of lower significance in economic terms. Nevertheless, there are individual topics which possess relatively high economic relevance. One example of this is a *fusion reactor*, the realisation of which however is highly uncertain and does not appear possible to the experts before the year 2025. (The spread for this estimate is very small!) The economic importance of this innovation will therefore probably not even be realised in the next generation. Named in second place with regard to economic relevance and importance for the labour market are *intense-field magnets based on high-temperature superconductvity*. Where this subject is concerned, relatively early realisation (median 2011) is combined with low statistical uncertainty in estimating the time-frame. From an ecological perspective, *chambers for climate simulation* and *geo-scientific investigations into the interactions of the oceans with the atmosphere* are assessed as being particularly relevant. In both cases, the hope of an alleviation of the global threats to the atmosphere is set alongside the goal of the pure acquisition of knowledge.

Figure 4.12-1:   What are the innovations in the field of Big Science Experiments important for?

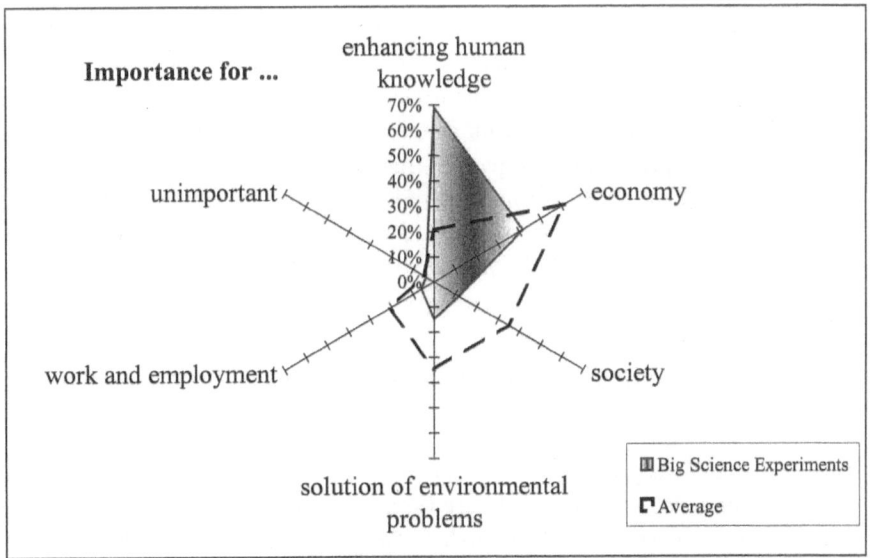

When will the topics become reality?

The strongly pure research-based character of the field of Big Science Experiments manifests itself in considerable uncertainty where time is concerned. The range in the distribution of the assessments for the time of realisation is only greater in the energy sector; in all other fields of inquiry of the Delphi study, however, it is considerably smaller. On average, the realisation times are five years further in the future than the average of the Delphi survey. Most individual topics fall into the five years from 2011-2015, whereas the corresponding cluster in the Delphi survey as a whole lies between 2006 and 2010. Only a single subject (the *development of neutrino detectors*) appears realistic as early as the year 2005.

What are the Utopian dreams of the field - those visions which cannot be expected to be realised by 2025 and perhaps never? The answer is unequivocal: the topics on *fusion research.* The experts grant the least prospects for realisation to a clarification of the question of *to what extent muon fusion reactors* are useful *for energy production or as a neutron source.* Almost 30 % do not believe this task is soluble at all. There are very many sceptics likewise when it comes to the question of whether the special *safety features of a fusion power station* can be *proved.* Even a scientific decision regarding whether *inertia fusion using heavy ion rays* is applicable for energy production is considered not to be realistic. On top of this the experts express objections in principle and regard for example *base load supply with electricity from fusion power stations* as unlikely in the next 50 years and refer to the continually negative energy balance and thus to the pointlessness of the development of a fusion reactor.

Another topic on the list classified as Utopian is *the production of a positron microscope* and the pipedream of *research reactors operated solely with non weapons-grade uranium.* At the same time, the experts consider both of these topics to be unimportant. The political dimension of the latter vision is expressed in the comments of some experts. In that respect references are made for example to the Munich research reactor under construction and thus to current political debates. This may have influenced the responses more than considerations of the future. Other experts declare the subject of the *fuel of research reactors* to be an "unscientific question".

What is the level of Germany's research and development for Big Science Experiments?

The German experts give big science equipment research in their own country good marks: in this field, more than 30 % see Germany behind the USA in second place and even assess other EU countries higher than Japan. In the parallel Delphi survey in Japan this is not corroborated. The experts there see the relationship between Japan and Europe in big science equipment research as exactly the reverse. The subject field is extremely international, at any rate European in nature, since more than 60 % of all those surveyed can no longer point to a national research picture but refer instead to cross-border common interests. This becomes especially clear in the case of fusion research: it is conducted almost entirely on an international basis. Even elementary particle and nuclear structure research and the area of major equipment as aids are judged from an international viewpoint. On the other hand, R&D in the fields of computational science and astronomy are still characterised more as national. Other countries of the European Union are classified by the German experts as particularly weak in the field of geo-experiments and computational science, Japan similarly in the field of astronomy and nuclear and elementary particle physics. The USA is regarded as markedly dominant in astronomy as well as computational science.

What needs to be done for Big Science Experiments?

Indications emerge from the Delphi survey with regard to bottlenecks in the field of science and technology policy in the broader sense. The most common mention by far concerns "international co-operation", its continuation and extension. Many big science experiments are barely conceivable any longer on a purely national basis. An improvement in the R&D infrastructure and support from outside funds are named in second place. In view of the discussion whether too much rather than too little big science is being carried out in Germany, this actually comes as something of a surprise. Big Science research will have a hard time in making profits from income under licence- and know-how agreements and suchlike, because there is no expectation of direct economic benefits, with the result that commercial sources of finance will largely be lost. Alongside institutional support, backing from outside funds is therefore especially important.

One extremely interesting aspect is the improvement in the exchange of personnel between industry and science. It is above all those employed by industry and to a lesser extent the scientists in major research themselves who call for improvements in the transfer of personnel. Accordingly, not even 7 % of the researchers questioned in state institutions, but almost 14 % of those representing industry, stress the necessity of improved personnel transfer. This, too, points to the fact that from the perspective of industry significant potential know-how exists which has obviously not yet been sufficiently recognised in the public research institute sector.

What follow-up problems might ensue?

A good many citizens who are critical of technology certainly do not deny the contribution of technology to the solution of a large number of mankind's problems, but at the same time point to the fact that it is double-edged: as current bottlenecks are removed, new problems simultaneously arise. Generally speaking, however, the field of Big Science Experiments is relatively "inconspicuous" when it comes to potential problems - how could it be otherwise where more discovery-led research is involved?

This on the whole minor fear of potential problems focuses on just a few questions. Most environmental problems are therefore expected to occur in the area of computational science, because both chip production and the disposal of computers have ecological dimensions. In this respect, the experts also view geo-experiments and nuclear fusion with scepticism.

Key point: Expansion of knowledge through big science experiments will become even more international

Despite the diversity of the individual subjects, in the subject field of Big Science Experiments, the Delphi survey contains a series of highly pointed clues to current trends. Research in the subject fields summarised here is already extremely Europeanised, i.e. cross-border in scope; it is no longer possible to point to a national research picture in nuclear fusion, nuclear structure and elementary particle research and in big science equipment for users. The interest of industry in this type of research extending beyond the direct application-based nature of the research results is unequivocal and certain. Along with this, the institutional boundary between science, big science research and industry is also becoming more open.

Improved interchange of personnel between industry and science is called for where the application-based visions are concerned, for example in *heavy ion accelerator plants with a broad range of services*, in *intense-field magnets based on high-temperature super-conduction* or in the use of *accelerators for the clinical irradiation of deep-seated tumours*. The experts express a wish for an improvement in the R&D infrastructure in relation to *accelerators for operating sub-critical nuclear power stations* and for *incinerating radioactive waste* and again for *heavy ion accelerator plants with a broad range of services* as well as *intense-field magnets*. The multiple mention of various technological policy measures stands out in the case of many topics. Thus, for the *use of heavy ion accelerators for tumour treatment*, more outside funds are also demanded by most experts and, where *accelerators for operating sub-critical nuclear power stations* and for *incinerating radioactive waste* are concerned, very many of them urge a change in regulations as a condition for realisation. It would also be desirable for the potential for co-operation across institutions in the field of big science research to be recognised and used more extensively by the public research institutes.

5   Megatrends: Stereotyped Thinking Pattern of the Experts?

5.1   Megatrends in Delphi '98 and their influence on science and technology

In general linguistic usage megatrends are understood to be social, political or economic developments (e.g. fashion, political preferences) which over a period of years move in a similar direction (e.g. increase in the figures in the annual statistical comparison). Generally speaking, the changes take place gradually and not suddenly. In order to detect them, at certain times in the social sciences the comparative values or characteristics determined in each case with specific variables are taken as the basis for the trend statement and can subsequently be used for further forecasts (Brockhaus Encyclopedia 1993, p. 343).

In each questionnaire of the first round of Delphi '98, 19 megatrends on global developments were brought up for discussion. These were selected from large international trend studies and answered with great interest by almost all those surveyed, with the result that a very sound assessment of more than 2,300 people exists on individual trends. First of all, those taking part had to assess whether they agreed with the megatrends in principle or not.

In the first round of the Delphi '98, the experts were additionally asked to reveal their personal opinion about the chances of the following megatrends and their probable influence on the general development of science and technology in the future (see chapter 2.3). At first it was asked for general agree- or disagreement (given in per cent). Table 5.1-1 describes the megatrends and gives their data. The time horizon was divided into seven five-year-intervals from "up to 2000" to "after 2025". Table 5.1-1 shows the percentiles (for calculation see chapter 2.5). The influence on the development of science and technology had four different categories on a Likert scale from "high", over "medium" and "little" to "none". For this, an index is calculated to weight ranging between 100 and zero.

Note that these 19 megatrends were added to each of the twelve disciplinary questionnaires in the same way. So every respondent had the same list of megatrends irrespective of her or his expertise and sectoral affiliation. Further, the megatrends were not iterated in the second round but sent out only once as the hypotheses of gaining stability of S&T experts´ assessments within their fields of expertise does

Table 5.1-1: List of 19 megatrends[4]

Megatrends	Number of the answers	Do not agree (%)	No opinion (%)	Agree (%)	Time			Influence of Science and Technology (Index)
					First Quartile	Median	Third Quartile	
1 World population will surpass the 10 billion (10 000 000 000) border	2301	19	9	72	2010	2017	after 2025	72
2 Low birth-rates and a constantly increasing life expectancy will in industrialised countries lead to over one-third of the total population being more than 60 years old	2306	7	4	89	2008	2013	2019	59
3 The globalisation of the economy will make national economic policy almost insignificant	2311	51	7	42	2005	2009	2015	74
4 Technical progress and the global reallocation of employment will increase permanently the medium unemployment rate in most of the developed countries	2302	22	4	74	1999	2002	2006	66
5 After reforms being realised, Germany will again become an internationally attractive location for investment	2278	27	13	61	2003	2005	2009	84
6 China's GDP per inhabitant will surpass that of the European Union	2309	56	16	28	2010	2016	after 2025	55
7 The Islamic nations are becoming the most dominant political alliance in the world	2310	67	16	17	2007	2012	2019	38
8 There will be violent conflicts between rich and poor countries	2303	56	14	30	2007	2011	2019	58
9 Massive migration will lead to riots in Germany	2299	49	14	37	2003	2007	2011	42
10 The European Union is going to develop a European government that will substitute national sovereignity	2305	42	7	52	2010	2015	2024	54
11 A world government is an effective institution for preventing and resolving violent conflicts	2309	76	7	16	2017	after 2025	after 2025	37
12 Climatic changes will lead to depopulation in wide regions	2309	48	15	37	2012	2021	after 2025	70

---

4   The index of the influence on science and technology is calculated from the weighted answers and ranges from zero to 100.

	Megatrends	Number of the answers	Do not agree (%)	No opinion (%)	Agree (%)	Time			Influence of Science and Technology (Index)
						First Quartile	Median	Third Quartile	
13	The world-wide scarcity of fossile fuels will enforce the rationing of energy consumption for private households	2308	41	6	54	2011	2018	after 2025	88
14	Most people in Germany do not found a family anymore	2304	71	13	16	2006	2011	2017	33
15	Tendencies of increasing individualisation and pluralisation hamper the functioning of the classic decision-making organs of representative democracies	2285	33	17	49	2003	2007	2012	52
16	Women will at least keep one-third of all executive positions in business	2305	32	11	57	2009	2014	2020	34
17	Due to low demand, more than half of the churches in Germany will be shut down	2305	40	18	42	2008	2013	2019	10
18	Technological development will give two thirds of all employees the opportunity of working at home	2306	62	7	31	2010	2016	2024	68
19	Increasing environmental problems will negatively affect the health of most people	2308	42	5	53	2003	2008	2015	79

not apply for personal opinions. Expressively, the respondents were advised to exclude their personal views when giving detailed S&T statements, but judging rather as citizens, and not as experts, when assessing the megatrends. More than 2.300 participants answered this part of the questionnaire, some were so enthusiastic, that they have left out the part on science and technology.

What do these opinions signify? We start with highlighting some descriptive results. (The actual wording of the surveyed theses is shown in italics.) According to the respondents, a continued *increase in the world's population* and an *ageing society in the industrialised nations* can be expected, at least in the intermediate future (see table 5.1-1). In the immediate future, there is the fear of a still greater and *permanent increase in the average unemployment rate in most developed* industrial nations. There is consensus in this regard. Apparently, the respondents do not feel that the political goal of halving the current unemployment rate can be achieved. In the intermediate future, however – and here again is consensus among those surveyed - *Germany can once again become an attractive location for investment*, provided that *reforms* are carried out.

A solid majority of the respondents believe – at least in the intermediate to long term future – in a *proportion of one-third of all leadership positions being filled by women*, a *world-wide shortage of fossil fuels resulting in the rationing of energy consumption by private households*, and that the *European Union will develop a European government that will supplant national sovereignty*. In the short-term, most of the respondents fear that *increasing environmental problems will negatively influence the health of most individuals*. Many abstained from responding to the question of whether the *trends towards increasing individualisation and pluralisation will influence the traditional decision-making processes of representative democracies*. Those who agreed with this megatrend expect such a development to occur in the first decade of the next century.

The experts' opinions were completely split when it came to the question of whether *more than half of the churches in Germany will close their doors due to lack of interest*. Those who felt this would occur, think it would more than likely be a long-term development. The rededication of church facilities might be seen as an alternative. Some examples for this alternative already exist in Germany.

"Globalisation" is the catchphrase on everyone's lips. However, less than half those surveyed believe that it will also *lead to national economic policies loosing practically all meaning*. Equally disputed are whether *massive migration flows will lead to unrest in Germany*, or whether *climate developments will lead to depopulation of large areas*. *Armed conflicts between rich and poor nations* appear even less probable. Also, there will probably not be any *world government* able to *effectively limit armed conflicts* - , not even in the long term. Once again, all those surveyed agreed on this latter point.

There is also skepticism regarding the question of whether *technical development will permit two-thirds of all employees to work at home*. Only a third of those surveyed believe this will occur. However, even they feel that this will not come about until the start of the second decade of the next century. Apparently, the question's requirement of "two-thirds" is set too high, even though the technical preconditions for "tele-working" are, to a certain extent, already in place today, and a further expansion is to be expected.

Neither do the respondents fear any economic or political dominance by foreign countries. Most believe that it is improbable *that the per capita gross national product of China will exceed that of the European Union* or that *Islam will result in the development of the world's strongest political block of nations*.

The family will also continue to play an important role in the lives of Germans; the question of whether *individuals in Germany will no longer start families* was resoundingly rejected. However, the received comments also pointed out that a "family" need not necessarily be linked to a marriage license, and that the understanding of what is meant by the term family was currently undergoing a transformation.

If the experts considered the respective trend to be correct, they were asked to determine the period in which it can become definite, and to judge to what extent the trend can influence science and technology (see table 5.1-1 and fig. 5.1-1). Accordingly, the *scarcity of fossil fuels and with it a rationing of energy consumption* could have the greatest influence on science and technology in the private sector. But here opinions are very divided: half of the experts fear precisely this, even if it does only occur in around 20 years or more, whereas the other half of the experts do not believe that things will get to that point. The following result is gratifying: by a large majority the experts think that *Germany can again become an attractive investment location* - though only after fundamental reforms. And this would have an immense influence on scientific and technical development.

Figure 5.1-1:   The six megatrends with the highest influence in science and technology

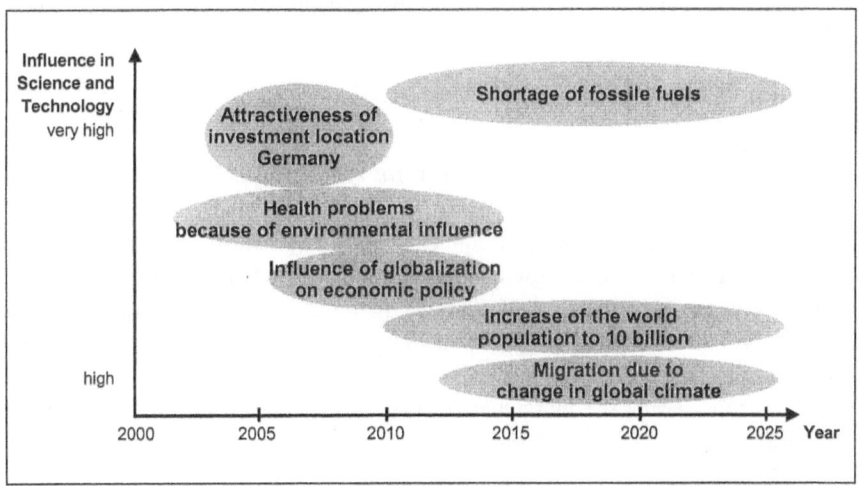

It again becomes clear from the results shown in figure 5.1-1 that it is above all those megatrends which are concerned with changing basic environmental conditions, i.e. the *increasing scarcity of fossil fuels, damage to health as a result of environmental influences* and *migration flows due to global climate changes*, which could have a very pronounced influence on the development of science and technology. Besides the environmental trends, extremely high political importance with regard to the development of science and technology is subordinated in particular to the regained *attractiveness of Germany as an investment location and the increasing influence of globalisation on economic policy*. This assessment shows clearly that with the aid of economic policy measures a thoroughly positive influence can be brought to bear on science and technology.

What is remarkable is that despite the vigorous public discussion, the evaluation on the economic rise of China and the political influence of Islam with respect to science and technology turns out to be not so high. Only 28 % of the respondents think that *China will surpass the European Union in per capita Gross Domestic Product* - and if so, only after the first decade of the next millennium. But: even if things perhaps do not reach that point, the growth of the Chinese economy must not be underestimated.

Still fewer experts believe that *Islam could develop into politically the strongest block of states in the world* (17 % of the 2,310 respondents) - and even these think this will be possible in 10 years at the earliest, and probably only much later. Finally, it should also be mentioned in this context that a considerable influence on the development of science and technology is also attributed to the continual *increase in world population*.

The remaining megatrends produce the following picture of opinions shown in table 5.1-1.

5.2 Factor analysis to determine R&D experts' attitudes

The megatrends in the Delphi questionnaires are the statistical base to arrive at the identification of the major societal forces driving and challenging S&T. Equally important is the identification of personal patterns in the answering behaviour, which would allow to classify the experts in research and development (R&D) into different clusters representing individuals with similar scenarios of the general future.[5] For this reason, the individual answers of the megatrends were used as the statistical basis for a factor analysis.

Factor analysis is a statistical technique used in psychology and in other social sciences to identify a relatively small number of factors that can be used to represent relationships among sets of many interrelated variables (Child 1975). For instance, in modern times, it is used to describe and identify attitudes toward religiosity (Kecskes/ Wolf 1993), toward foreign language (Corbin/ Chiachiere 1995), toward reproduction (Wikman et al. 1992) or ideology (Hayduck et al. 1995). As far as we know, it has not yet been used to describe attitudes of R&D experts toward general

---

5   Demographic measures may also be interesting to include into the analysis, however for the factor analysis approach the very unequal distribution of the experts by gender (less than 10 % females) does not promise signficant results.

future megatrends. Our aim is to find typical patterns in the response of the experts to the megatrends in order to identify typical schools of thought about the future.

In terms of mathematical statistics, factor analysis is a "scoping" method which does not test for the significance of postulated hypotheses, but it rather tries to find few "latent" variables among many, not linearly independent ones. It turns out as a result of factor analysis if there are any (Kaiser criterion), and if so how many such latent variables (factors) exist and to what extent these explain the variances between the cases (respondents).

Because of the problem of missing values[6], every missing and every "no opinion" answer was replaced by the value of the median. This was done to get factor scores for every individual to be able to identify his or her loading on the different factors. The answers of the attitudes towards the realisation of the megatrends were converted into a modified 8-point Likert-type scale, ranging from realisable until 2000, over the five five-year-intervals to general disagreement with the megatrends.[7]

Principal components analysis was performed with varimax rotation. The statistical program used in this study was SPSS factor analysis (Norusis 1993). Five factors with eigenvalues greater than 1.00 were found and confirmed by means of a scree chart (Catell 1978). First, a factor analysis was run with all 19 megatrends. A subsequent neglection of the megatrends no. 6 *(China's GDP per inhabitant will surpass that of the European Union.)*, no. 7 *(The Islamic nations are becoming the most dominant political alliance in the world)*, and no. 11 *(A world government is an effective institution for preventing and resolving violent conflicts. For the numbers see table 5.1-1.)* did not turn out to increase the significance of the factor analysis, because the great majority of the respondents disagreed with these statements, and did not specify information on the time scale, therefore.

Bartlett's test of sphericity was significant (3130.06; $p < 0,001$) suggesting that the correlation matrix among the remaining 16 megatrends was not an identity matrix. The Kaiser-Meyer-Olkin measure of sample adequacy reached a value of 0.753. Kaiser (Kaiser 1974) characterises measures in the 0.70's as middling, therefore we can comfortably proceed with the factor analysis.

---

6   Only around one third of the over 2.300 respondents answered every question without using "no opinion".

7   In this way, the factor analysis was not only based on the answers of the categories "do not agree", "no opinion" and "do agree" which would have led to a more rigid separation of the factors by neglecting also the interesting information of the highly variant answers on the expected time of realisation. The respective factor analysis based only on agreement, but not on the time scale, resulted in similar factor patterns. In the Austrian Delphi, the latter option was used.

Often the variables (here: megatrend items) and factors do not appear correlated in any interpretable pattern. Most factors are correlated with many variables. Since one of the goals of factor analysis is to identify factors that are substantively meaningful, the rotation phase of factor analysis attempts to transform the initial matrix into one that is easier to interpret. Although rotation does not affect the goodness of fit of a factor solution, the percentage of variance accounted for by each factor does change. The most commonly used method is the varimax method, which attempts to minimise the number of variables that have high loadings on a factor. This should enhance the interpretability of the factors. The factor loadings are published in Blind/ Cuhls/ Grupp 1998 and shall not be repeated, here.

Factor 1 is the dominant factor, accounting for 1´ 1 % of the variance of the total factor solution, with five megatrends attaining loa ings from .45 to .64. The megatrends of Factor 1 are concerned with an optimis c view towards social problems especially in Germany but also between poor and i ch nations in general.

Factor 2 accounts for 9.0 % of the variance in the ι tal solution and has three highly loaded megatrends. The factor expresses a sceptic. attitude towards both *political progress in Europe,* the *emancipation of women in profession,* and the *broad diffusion of tele-working.*

Factor 3 accounts for 7.4 % of the variance of the total factor solution. The three highest loading items of the factor express an optimistic attitude towards our *environmental and resource problems.*

Factor 4 and factor 5 explain 6.9 % and 6.4 % of the variance "respectively" in the total solution. Megatrends 1 *(World population will surpass the 10 billion border.)* and 2 *(Low birth-rates and a constantly increasing life expectancy will in industrialised countries lead to over one-third of the total population being more than 60 years old.)* load especially heavily on Factor 4, which expresses an optimistic attitude towards a sustainable development of the population in the developed and less developed countries.

Factor 5 was the least important of the interpretable dimensions, with three medium-loaded megatrends concerning the effectiveness of national "location"-policy in the age of globalisation, especially with an optimistic view for Germany after implementing serious reforms.

In total, only 46.8 % of the variance is explained by the five factors, this means that there exist further hidden dimensions, which are not covered by the 16 megatrends. Therefore, in future surveys some additional dimensions, like demographic factors, which could describe probably more personal attitudes towards the future should be included. However, in a similar exercise this general pattern of factors is confirmed, but no better values are reached (Bundesministerium für Wissenschaft und Verkehr 1998).

Since one of the goals of factor analysis is to reduce a large number of variables to a smaller number of factors, it is often desirable to estimate factor scores for each case. The factor scores can be used in subsequent analyses to represent the values of the factors. Because of the five factors, we get five dimensions. One usually inverts the factor matrix in order to estimate factor scores for each case (person). With admitting combinations of dimensions, we would have identified $2^5 = 32$ groups of experts, one explained by the five factors straight forward (positive factor scores), one being positioned diametrical to the five factors (negative factor scores). The other 30 represent mixtures of positive and negative scores per factor. This is the same as the distribution of data in four quadrants in case of two factors.

There are several methods for estimating factor score co-efficients. Each has different properties and results in different scores (Tucker 1971; Harman 1967). The three methods available in the SPSS Factor Analysis procedure (Anderson-Rubin, regression, and Bartlett) result in the same factor scores, because principal component extraction is used. This means, in our case, there is no need to estimate the factor scores, but they can be calculated unambiguously.

For facilitating the further evaluation of the detailed S&T assessments, we prefer to code the experts one by one, i.e., in a disjoint manner, to arrive at a similar "factor cohort" code as, for instance, the age cohorts. In doing so, the calculated factor scores can be used to attach to every expert only the single factor where she or he has the highest value or where there are diametrical attitudes in case of very negative values. The result are five pairs or ten groups ranging from (+1,-1) to (+5,-5). In this case, an eleventh group (0) of so-called mixed experts should be defined for those persons who do not have high scores in any dimension. In our case, factor scores between -.5 and +.5, which do not allow definite attachment to one factor, were taken as the threshold. The result of this grouping gives the distribution in table 5.2-1.

Table 5.2-1:   First-stage grouping of R&D experts

Groups	+5	-5	+4	-4	+3	-3	+2	-2	+1	-1	0
Percentage	9.5	8.5	9.9	8.4	4.6	10.4	6.2	9.4	6.8	7.7	18.6

The group of mixed experts is the biggest with more than 18 %, whereas the persons explaining the factor of an optimistic view towards the environmental conditions represent the smallest share with around 4.6 %.

If we remember our original aim and the data on the estimation on the future of science and technology, it turns out that we cannot continue with eleven subgroups between 113 and 456 experts, because we have to bear in mind to divide these numbers by twelve questionnaires. Consequently, the number of respondents per questionnaire respectively per question would fall beyond a statistical significant sample size of around 30 not allowing a significant comparison. Therefore, the eleven groups are sensibly regrouped into four equally sized expert-types plus the group of mixed experts being aware of the inaccuracy of this approach. In other studies an even more rigid approach, only optimistic and pessimistic types, and mixed persons are divided in three groups because of the same data restrictions (Bundesministerium für Wissenschaft und Verkehr 1998).

The experts type 1 are those individuals, who have positive factor scores either in the Factor 1, Factor 3, or Factor 5, therefore explaining these three factors. This kind of person is in general very optimistic or has an optimistic scenario in personal mind, especially concerning the economic and social problems of Germany or the global environmental and economic problems. Therefore, we call this expert shortly the *"location Germany optimist"*.

The experts type 2 are those respondents, who have negative factor scores in the Factor 3 or in the Factor 4 and are therefore not contributing to the explanation of these factors but even lying diametrically to them. They are pessimistic either towards the sustainable development of the population growth or towards the melioration of our natural environment. In abbreviation, we speak in the remainder about the *"environment pessimist"*.

The expert group type 3 consists of individuals, who have either positive factor scores of Factor 2, contributing to his explanation, or negative factor scores in the Factors 1 and 5, respectively being diametrical to the latter. These persons are very sceptical towards technical, social and political progress in general or are looking into the future of Germany less positively. In short form, we name this group the *"progress sceptics"*.

The expert group type 4 is defined as those persons, who have either a positive factor score or explanatory power of Factor 4 or being diametrical towards Factor 2. He or she is characterised as someone who is convinced that the population problem in the world and the overaging in the industrialised countries can be dealt with successfully or believes in the technical, social, and political progress. This character will be called the *"population optimist"*.

The remaining persons of expert type 5 are the so-called mixed experts, who have no significant, exceptional opinion in the before-mentioned issues.

Figure 5.2-1 shows the relative distribution of the expert types in the twelve fields of the Delphi ′98 study, who have answered in the first round. Concerning the relative distribution of the expert types in the twelve fields of the Delphi ′98 study, who have answered in the first round, in general, every group consists of around one fifth of the total. However, there are differences between the fields. A chi-square test of independence of two variables, here the five types and the twelve fields, underlines that there is dependence between these two variables. Especially, the over 30 % environmental pessimists in the field Space are significantly different from the expected value. Furthermore, the 27 % population optimists in Services and Consumption are outside the statistically expected range. Before we come to the estimation of technical questions, we are able to conclude that at least the distribution of the different types in the twelve fields differs significantly from a random one.

Figure 5.2-1:   Distribution of optimists, pessimists and neutral persons in the individual subject areas (in %)

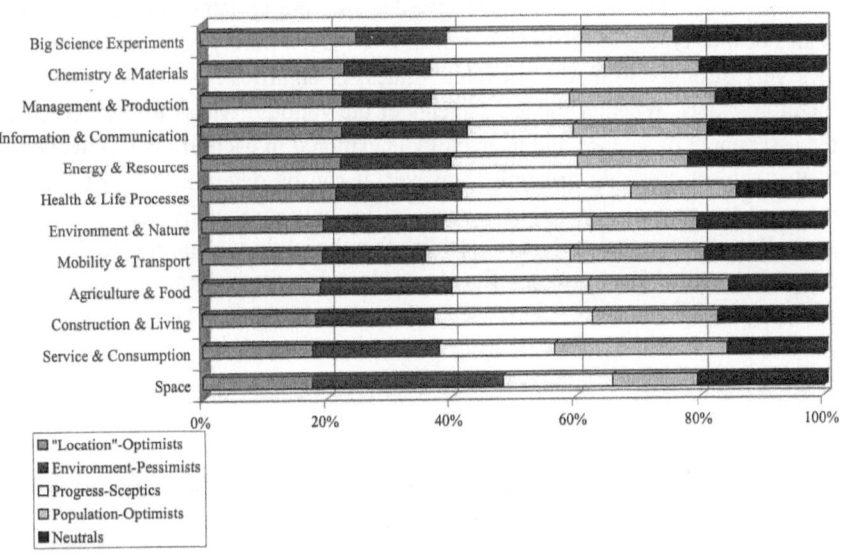

Overall, the group of optimists and the group of sceptics each with 40 % (20 % being neutrals, i.e. people responding in mixed fashion) roughly balance each other out. Conspicuous features can be discovered if one looks at the composition of the experts in the individual subject areas on a somewhat more sophisticated basis. Thus, in the subject area of Chemistry and Materials, the *location optimists*, who no doubt remember the international competitiveness of the German chemical industry, enjoy above-average representation. Many experts in Management and Production are extremely optimistic when it comes to the solution of national and global population problems and thus also see an opportunity to be able to provide people around the world with an adequate supply of goods. An above-average number of *environmental pessimists*, but also *population optimists*, can be found among the experts on Agriculture and Food. The first group is worried about impending environmental problems, particularly in the context of resource-intensive production in agriculture, while the second group believes that in the event of sustained population growth the world-wide food problem can be solved.

The *location optimists* are extremely well represented in the areas of Energy and Resources and Big Science Experiments. Among the energy experts and the staff of the large research institutes the conviction therefore prevails that in the foreseeable future Germany, if the appropriate structural changes take place, can again become an attractive investment location. Finally, the *population* and *location optimists* are also the largest groups among the experts on Information and Communication as well as on Services and Consumption who, against the background of the current trend towards the information and service society, in the majority back the accomplishment of further fundamental technical and social innovations.

On the whole, statistical tests have revealed that the variable composition of the shares of the "types" in the twelve subject areas is not purely coincidental but deviates very strongly from the expected values. A significant statistical connection therefore exists. Whether the assessments of the various "types" find expression in the assessment of science and technology is explored in each of the chapters on the twelve subject areas. In principle, however, it can be determined that the assessments of the various "types" on average differ only marginally across all subject areas. It is impossible therefore to establish that the individual attitude to the general future of mankind has a strong influence on the subjective assessments of the future of science and technology.

5.3 Similarities and differences in the estimation of the future development of science and technology

The following results are based on the average answers on all 1.070 detailed statements about the development of S&T, and the anwers of the five expert types on the different categories, which were asked for.

Expertise

As explained, the experts had to assess their degree of expertise for every statement. The general distribution of the self-estimated expertise of the experts into the three categories "high expertise", "medium expertise" and "low expertise" is very similar between the five expert types. However, the *"location Germany"* *optimists* have a higher share of specialists in their group, whereas the *population optimists* claim relatively often only a small expertise.

Importance

Reacting on the critique on the first Delphi study of being too imprecise by asking for the overall importance of a topic, this category was split into importance for "the enlargement of human knowledge", for "the general economic development", "the development of the society", "the solution of environmental problems" and finally for "labour and employment" (chapter 2.3ff, fig. 5.3-1). In general, the "location Germany" optimists assess the asked visionary themes as more important, especially for work and employment. Consequently, both the environment pessimists and the population optimists stress the importance of the topics for the solution of environmental problems. There is a mixed emphasise on the importance of the ideas for the economy. Contrary, the progress sceptics have a more reluctant view towards the different importance categories and assess more of the proposed innovations as being not important. We have to infer that the importance reactions depend on personal characteristics of the experts.

Figure 5.3-1:   Dimensions of importance

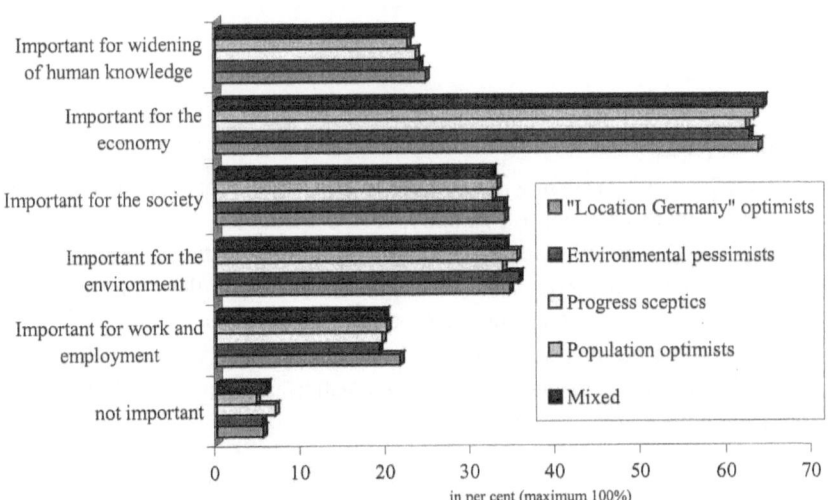

Estimation of the realisation time

The time horizon enjoys the highest attractivity among the various results of Delphi reports. The time of realisation divided into seven five-year intervals was asked for, as in the previous German and Japanese surveys. The range of the average time realisation of all 1.070 topics between the five groups is surprisingly small (fig. 5.3-2). It is not even one year. Therefore, we can conclude that the highly aggregated expectation of the realisation time does not depend significantly and consequently on the personal attitude towards megatrends.

Figure 5.3-2:   Expected time of realisation

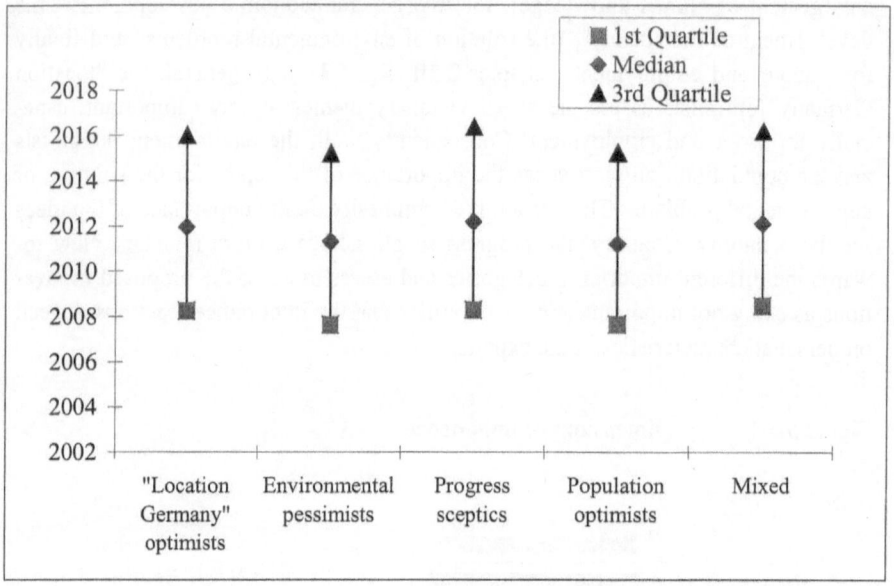

Leading country in R&D

To find out the relative standing in R&D, among the USA, Japan, Germany, another EU country and another non-EU country, it was asked for the leading nation in R&D (fig. 5.3-3). For all expert types, the USA are the leading country in R&D. The *"location Germany" optimists* tend to give more, while the progress sceptics tend to give less answers than the three other groups. Maybe, the latter ones are more in doubt and the first ones are more certain about the situation in R&D. The *population optimists* are favouring the R&D situation in Germany slightly more in comparison to the others.

Figure 5.3-3:   R&D level

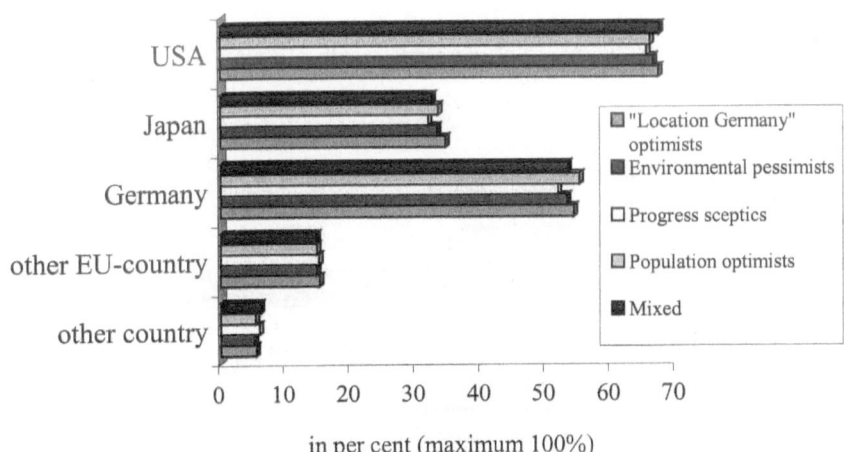

in per cent (maximum 100%)

Policy measures

In order to get closer ties to S&T policy, one prominent new part of the assessment criteria were the kinds of measures - instead of possible obstacles in preceding studies - that should be taken to improve the situation (see chapter 2):

- *better education and qualification of scientists and technical personnel,*
- *exchange of personnel between university and industry,*
- *international co-operation on project level or for mutual knowledge and personnel exchange,*
- *an improvement of the R&D infrastructure,* e. g. the foundation of institutes, data bases or providing venture capital,
- *support by third parties* (state, foundations) etc., e. g. more financing of lead projects ("Leitprojekte"), immaterial measures, or
- *a change of regulation*: This can be deregulation, strengthening existing regulation, re-regulation (new regulation) or other changes in the national frame conditions (laws, norms, decrees, technical guidance, charges etc.).

Overall, there is a broad agreement about the further internationalisation of the R&D co-operation (fig. 5.3-4), especially the *environment pessimists* have a great confidence in this strategy. The *progress sceptics* are more hesitant in their support to globalisation and other strategies, but favour changes in the regulatory framework and a better education. The *"location Germany" optimists* prefer highly the support by third public or private parties or foundations and the improvement of the

R&D infrastructure. In general, the dimension of different policy strategies give hints that the different type of experts have different views about adequate policy instruments to support the realisation of innovations and new ideas.

Figure 5.3-4:   Important measures

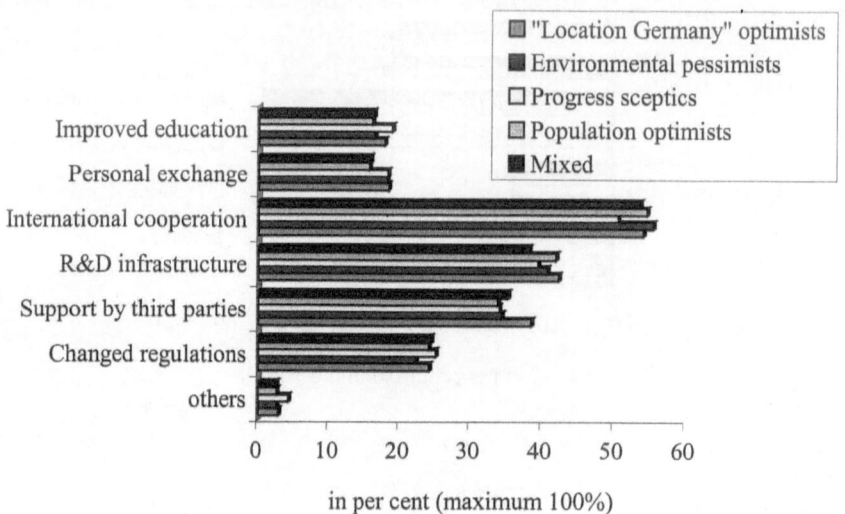

in per cent (maximum 100%)

Follow-up problems

In order to integrate aspects of the traditional technology assessment, it was asked for follow-up problems for the environment, for personal safety and for society or culture, in case the statements will be realised. The results of the follow-up problems do not show any significant and reasonably interpretable connection between expert groups and answering behaviour (fig. 5.3-5). However, both the slightly higher fear of problems in the environment by the population as well as *the "location Germany" optimists* and the lower ranking of social problems by the *progress sceptics* are surprising.

Figure 5.3-5:   Follow-up problems

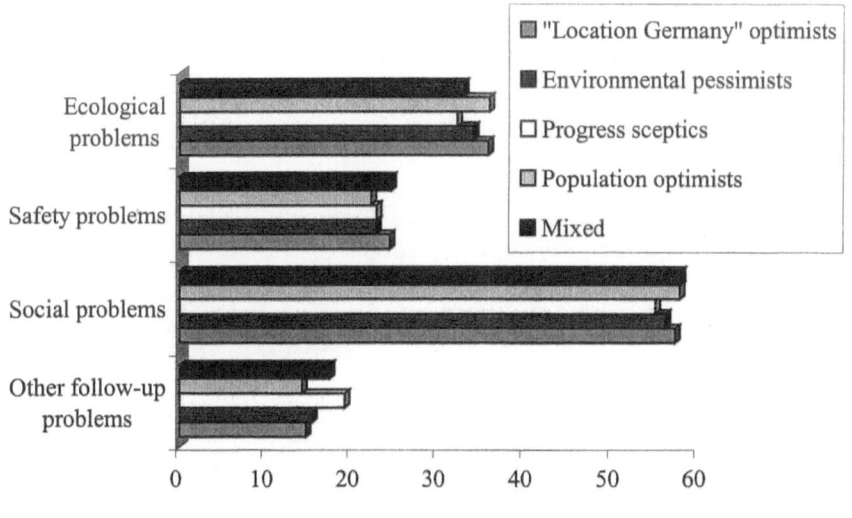

in per cent (maximum 100%)

5.4   Different estimations concerning innovation fields –
A selection

The comparison of the aggregated results of 1,070 statements between the five expert groups veils significant differences in single topics. On the level of single innovation fields, differences occur only in some cases (see e.g. chapter 4.8). In general, therefore, it can be stated that the influence of the "world view" of the experts has a marginal effect on their technical assessments. They assessed the single topics and their content rather than just judging one-sidedly as a "pessimist" or "optimist". But as some differences occur, they should nevertheless be mentioned here.

The most interesting criteria in this respect are the importance, the measures to be adopted, and the follow-up problems. For the *time of realisation*, there are no significant findings: in some cases (e.g. in Big Science Experiments), *the "location Germany" optimists* are even estimating later realisation times than the *pessimists* in general.

For the *importance*, there are also no differences between experts who judged more optimistically on the megatrends than those who were more pessimistic. In some fields like Environment and Nature, there is a rather mixed picture on the importance, with the *population optimists* assessing a lower importance than the others (fig. 5.4-1).

Figure 5.4-1:   Importance in the field Environment and Nature (in %)

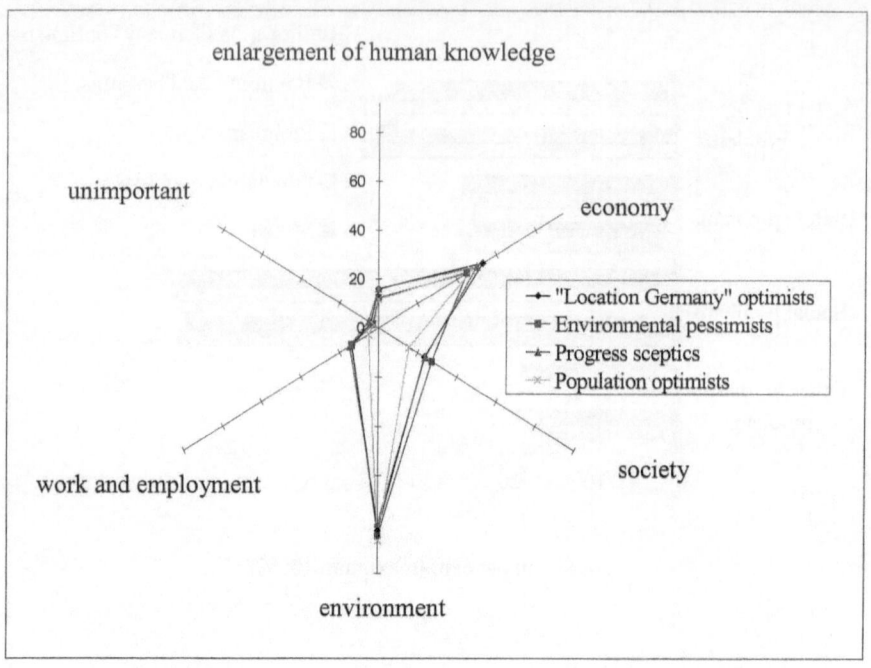

Figure 5.4-2:   Measures to be taken in the field Construction and Living (in %)

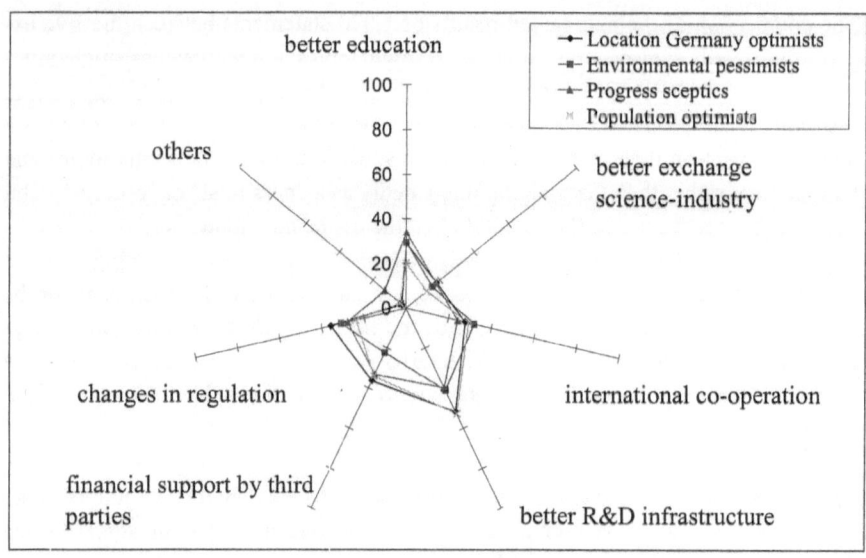

Figure 5.4-3:   Measures to be taken in the field Big Science Experiments
                 (in %)

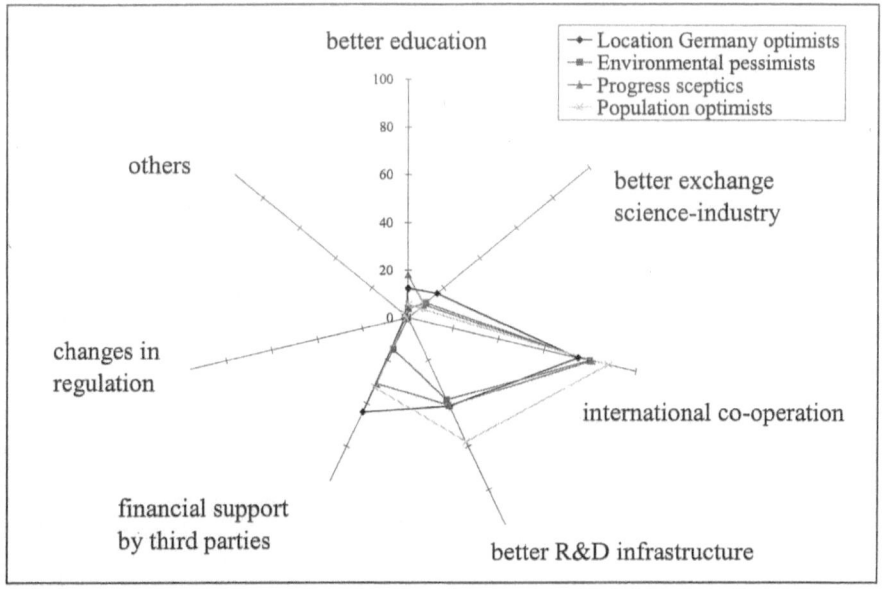

Regarding the *measures* to be taken, there is a mixed picture with slight differences to be observed in the fields Construction and Living, and Big Science Experiments (see fig. 5.4-2 and 5.4-3) as well as in Space.

For Construction and Living, the *progress sceptics* and the *population optimists* regard a better R&D infrastructure as necessary. Together with *the "location Germany" optimists*, they think that more financial support is also necessary. Only *the environmental pessimists* do not ask for financial support very often. That makes sense in a logical way, but does not explain why the other measures are not mentioned more often by these groups of persons.

In the case of the Big Science Experiments, the population optimists argue for more international co-operation, a better R&D infrastructure and more financial support by third parties. For the last mentioned category, only the *"location Germany"* optimists mention this measure more often, whereas the environmental pessimists do not ask for more measures in this respect. Big Science Experiments is a field in which most of the topics concern basic research, which is more dependent on public money than other fields. This might be a reason why these measures are asked for so often in general, but does not explain the differences among the groups of persons.

Figure 5.4-4:   Follow-up problems in the field of Space (in %)

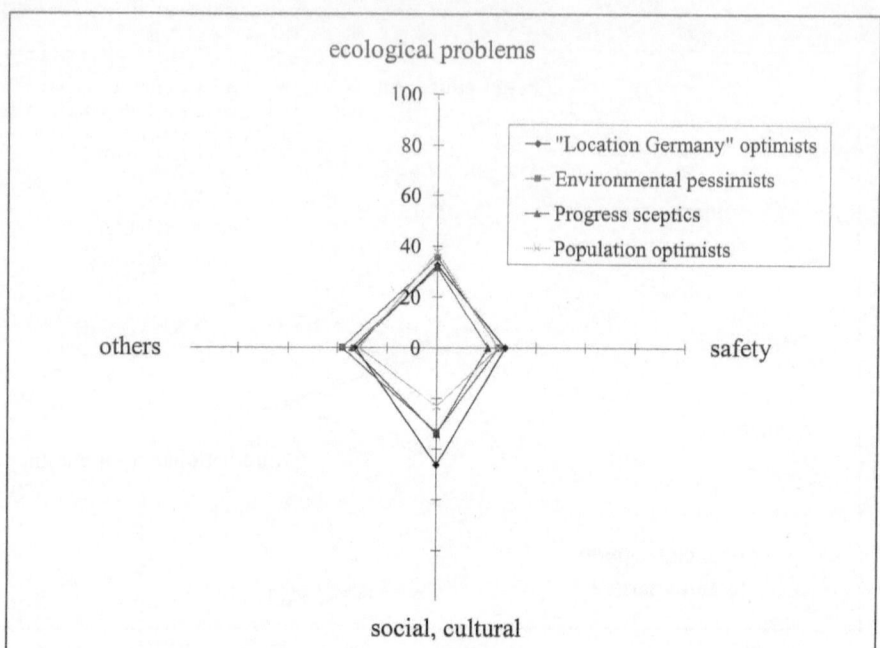

The hypothesis concerning the *follow-up problems* was that the *environmental pessimists* will generally expect more ecological problems. In the previous section, it was explained that overall this was not found to be the case. Looking at single fields, however, the differences are so slight that a relationship cannot be stated either.

The largest gap concerning measures can be found in the field Space (fig. 5.4-4), where the population optimists expect more social-cultural problems than the other groups and the *location Germany optimists* expect less problems. Especially in the field of Space, the optimists expect rather *more* problems than the sceptics. There are just slight differences in the expectation of safety problems, so that no significant difference in the answer behaviour can be found even in this field.

5.5   Selected topics with significantly different estimations by the expert types

Some differences occur, however, do occur on the level of single topics. In the next section, some examples will be discussed provided that at least 100 experts in total, respectively around twenty of each group, have answered. This means that we leave out topics with a low response rate in order to reduce the statistical problems of randomness of small samples. Without this restriction, more topics could be found.

The topic *cars are more compact and much lighter than 1.000 kg in total* is assessed as important for the economy by more than 80 % of the *environmental pessimists*, whereas less than 40 % of the mixed and around 45 % of the *progress sceptics* are convinced of the economic importance of this statement. Concerning the second topic, *radar systems for the prohibition of car accidents are widely used,* over 80 % of the *population optimists* consider this innovation of economic importance, but only 40 % of the mixed ones. The medical innovation *with the help of great DNA and protein sequence databases and advanced software programs it is possible to identify the protein function from the genetical sequence* is ascribed economic importance by over 70 % of the mixed experts, whereas only 30 % of the *population optimists* attribute it with an economic dimension. However, the main explanation for this difference lies probably in the 25 % share of high-level experts of the mixed group and the 0 % share of experts with highest expertise of the *population optimists*.

As regards the time frame, the following issues are controversially assessed by the expert types. For the topic *a communicative system with artificial intelligence is widely used, which gives impulses for the generating of ideas or detects disguised connections,* the median of the environmental pessimists hope that a realisation takes place around 2010, while the mixed person/group sees this statement realised after 2020, a difference of more than ten years. Again, towards the topic *medicine drugs, so-called "artificial angels", which stimulate the human joy centre and substitute drugs, are widely used* the median of the *environmental pessimists* has an expectation of around 2013, whereas the progress sceptics do not expect this vision to become true before 2024.

Finally, the topic *an "electronic parliament" acts in TV programs in order to decide about laws with the help of electronic polls* is puzzling. Surprisingly, over 50 % of the progress sceptics expect a realisation around 2016, whereas the other 50 % do not believe in a realisation at all. Furthermore, only 30 % of the *"location Germany" optimists* assume a realisation before 2025, while the rest deny that this vision will ever come true.

This small selection of topics underlines that for single issues personal visions about the world in the future influence the assessment of technical issues, either that the technology is contributing to the fulfilment of the general vision or in the way that it is hopefully prohibiting an expected negative scenario. Generally, more optimistic persons will also be those who are motivated to realise things.

We have seen that the R&D experts asked in the Delphi '98 study have different views on the general future of the world. There are characters who are rather optimistic about the future of the German economy and the possibility to use national economic policy effectively in the age of globalisation. Another group is confident about a sustainable development of the population, at least in the next 25 years. On the other hand, a relatively large share of the experts is very pessimistic about our environment and the supply of natural resources, and a fourth type is rather pessimistic about political, social and technical progress in general. The remaining persons have no significant answering behaviour.

Nevertheless, their estimation about the future of science and technology does not differ so much, if one looks on the aggregate results of all 1,070 topics. On the aggregate level, few significant differences can be identified and these occur more in the assessment of importance (i.e. priorities) and the evaluation of policy measures, but not in the time scale. However, the analysis of single topics brings out different answering patterns. If the development of science and technology is viewed in an integrated framework of the environmental, social and political systems both in a global and national dimension, then it is rational that differing perceptions of these conditions influence the perceived demand and importance, and therefore also the progress of new technologies.

For the reliability and validity of the Delphi method in general, it turns out that in general the influence of personal views about the megatrends of the respondents on the assessment of technical questions is only small. Though the opinions about the policy measures which should be adopted are significantly influenced by personal attitudes towards the future in general. The obvious dependence of the assessment of selected technical visions on different perceptions of general framework conditions supports the consistency of the personal answers and is not critical for the Delphi method itself. In a review of an international comparison of technology foresight studies, it is also emphasised "that social context matters" (Klein 1998). However, these results support the need of sufficiently large sample sizes of R&D experts to diminish the influence of respondents with extreme perceptions of general framework conditions. Furthermore, the integration of questions regarding the circumstances of a specific research topic is helpful to detect hidden links between the social, political and environmental framework and the possible development of a technology.

6   Comparison of the Asessments between Delphi '98 and Former Studies

Over 100 topics from the first German Delphi of 1993 survived the critical selection process by the technical committees designed to help ensure their topicality and these items were subsequently queried in an identical manner in the questionnaires of the second Delphi study. A comparison of the results comes directly to mind, because one instinctively asks oneself whether the experts of today share the assessments of the experts questioned approximately five years ago or not.

Table 6-1:   Topics with more pessimistic prospects of realisation

Delphi '98 field	Topic	Median Delphi '93	Median Delphi '98	Diffe-rence
INFOR-MATION	After clarifying the physiological and psychological connections on which abnormal human behaviour is based, a system which warns people of defects in good time is put into practical operation.	2009	2024	15
ENERGY	A heating and cooling system with a heat pump based on the use of solar energy is in practical use.	1999	2011	13
ENERGY	An FBR system (fast breeder reactor) which is integrated into a nuclear fuel cycle comes into practical use.[8]	2014	after 2025	12
CHE-MISTRY	Multilayer solar cells which attain an energy conversion efficiency of more than 50% are in practical use.	2008	2020	12
ENERGY	A technique for coal hydrogenation is in economic operation.	2005	2017	12
SPACE	Multiple-rendezvous exploratory probes are sent into interplanetary space and to the minor planets.	2004	2015	12

The estimations of that time have been shifted still further into the future, principally in the subject field Energy and Resources (table 6-2), but also in the field of Chemistry and Materials (table 6-3) and Space (table 6-4).

With regard to the expected time of realisation, the expectations expressed at that time have shifted on average by more than five years (median) into the future.

---

8   For this case, see also Cuhls 1998.

Topics are adopted into this mean value with authoritative importance which are now expected more than ten years later. Since this basic tendency may spring from a psychological effect or result from the fact that many of the themes which were previously "likely" are no longer brought up for discussion for reasons of relevance, table 6-1 merely shows the topics projected way into the future.

Table 6-2: Topics in the field of Energy and Resources with a shift in the expected time of realisation

Energy & Resources	Delphi '98 time of realisation			Delphi '93 time of realisation			
	Q1	Median	Q2	Q1	Median	Q2	(in %)
An FBR system (fast breeder reactor) which is integrated into a nuclear fuel cycle comes into practical use.	2007	2013	2018	2019	after 2025	after 2025	26,4
A nuclear fusion reactor is developed and operational. The problems with nuclear waste disposal are solved.	2017	2021	2023	2027	after 2025	after 2025	27,9
The biological production of hydrogen utilizing solar energy by means of intact organisms or biological sub-systems is in industrial application.	2009	2013	2019	2018	2023	after 2025	14,4
The temperature differences in sea water are in practical use to produce electricity.	2010	2017	2022	2019	after 2025	after 2025	34,2
A technique for coal hydrogenation is in economic operation.	2000	2004	2010	2011	2017	2023	3,4
Supra-conducting energy storage systems with capacities similar to those of pumped storage power stations (one million kWh) are in practical use.	2012	2017	2022	2021	after 2025	after 2025	18,2
A heating and cooling system with a heat pump based on the use of solar energy is in practical use.	1995	1998	2001	2007	2011	2018	4,3

Table 6-3:   Topics in the field of Chemistry and Materials with a shift in the expected time of realisation

Chemistry & Materials	Delphi '98 time of realisation			Delphi '93 time of realisation			
	Q1	Median	Q2	Q1	Median	Q2	(in %)
Organic, supra-conducting materials are developed whose critical temperature lies over 77K.	2003	2007	2011	2012	2017	2020	6
Organic materials which turn luminescent when supplied with energy and emit coloured light for over 3,000 hours, are in practical use.	2001	2005	2008	2007	2011	2017	0
High-performance building elements are developed for which materials with non-linear optical effects of the third order are used.	1999	2002	2004	2007	2010	2014	1
Synthetic membranes, which have an active transport mechanism analogous to biological membranes come into practical use.	2003	2007	2020	2009	2014	2018	0,8
A supra-conducting substance is developed whose transition temperature is normal temperature.	2007	2013	2019	2015	2022	after 2025	21
Irrespective of the type of material, the hetero-epitaxy technique is developed on silicon crystal wafers.	1997	1999	2005	2008	2010	2014	3,3
According to a technique used with scanning tunnel microscopes (STM), methods are developed to regenerate surface flaws on silicon crystal wafers and seal impurities.	2001	2004	2009	2006	2010	2014	1,9
Through the development of a steel glue with high adhesive and performance properties, construction with steel is considerably rationalised.	2002	2005	2008	2008	2011	2017	7,3
Light wave guide materials of greater lengths for transmission rates of 100 Gb/s are developed (at present 20~30 Gb/s).	1998	2001	2004	2006	2009	2013	0
A technology will be developed by means of which extensive semi-conductor mono-crystal layers can be produced on glass.	1999	2004	2008	2010	2014	2018	4,3

Multilayer solar cells which attain an energy conversion efficiency of more than 50 % are in practical use.	2003	2007	2010	2014	2020	after 2025	14
A process which dissociates water by means of solar radiation is in practical use.	2007	2013	2019	2013	2018	2024	8,1
Large-scale amorphous silicon solar cells which attain an energy conversion efficiency of more than 20% are in practical use.	2001	2004	2007	2009	2013	2017	3,3
A method to immobilize carbon dioxide comes into practical use, with which the CO_2 emissions are reduced to half.	2006	2011	2020	2014	2021	after 2025	28
The percentage of hydrogen-powered cars which are equipped with hydrogen-storing metal alloys amounts to over 10%.	2006	2010	2015	2016	2019	2024	16
A new refining technology is developed which makes mining titanium as cheap as aluminium.	2003	2006	2012	2010	2014	2018	21

Only in the case of the following three topics could a somewhat more optimistic assessment of the present-day experts compared with that time be established:

- *Effective methods for cultivating functioning eco-systems in former tropical rainforest areas are developed.*

- *Intelligent robots are in use which have sight, hearing and other sensory functions, can assess the situation in the outside world for themselves and make decisions independently.*

- *High-frequency broadband amplifiers in semiconductor technology in the frequency range above 1000 Ghz are in practical use.*

However, it must be remembered that topics realised in the meantime were not asked in the second Delphi and among these there are many which have come true earlier, including the widespread dissemination of "pocket telephones". Solely due to this selective effect, it was not possible to expect an estimate which is identical on average.

Table 6-4:   Topics in the field of Space with a shift in the expected time of realisation

Space	Delphi '98 time of realisation			Delphi '93 time of realisation			
	Q1	Median	Q2	Q1	Median	Q2	(in %)
A manned spacecraft lands on Mars and returns to earth.	2016	2019	2022	2021	after 2025	after 2025	6,5
A manned research station is established on Mars.	2014	2016	2018	2027	after 2025	after 2025	10
Multiple-rendezvous exploratory probes are sent into interplanetary space and to the minor planets.	2000	2003	2006	2010	2015	2020	0
Solar power stations with extra-large solar cell panels are set up in space whose energy is transmitted to earth in the form of microwaves.	2011	2014	2019	2019	2025	after 2025	16
Following the tethered satellite method, a satellite will be tethered to a space station and used to alter the conditions of gravity, produce electricity and accelerate the payload.	1997	2001	2007	2009	2012	2015	4,1
High-performance space transporters to transport large structures between orbits near to earth and the geostationary orbit are developed.	2004	2008	2014	2014	2019	2025	1,9

A further question which comes to mind focuses on the shifting of precedences and importance ratings. Whereas in the first Delphi study overall questions were asked about importance using four different grading possibilities ranging from "very important" to "unimportant", in the second Delphi there are five different dimensions of importance which were added to produce a basis of comparison. The list of the topics whose importance has risen sharply compared with the first Delphi study – based on a ranking analysis – makes it clear that in particular information and communication technology and its applications in the educational sector have gained still further in importance.

- *Intelligent robots are in use which have sight, hearing and other sensory functions, can assess the situation in the outside world for themselves and make decisions independently.*

- *Bio-analog materials with perception and decision-making capabilities are developed.*

- *Systems are developed which are able automatically to produce summaries or extracts from books or documents (print-out in any length required).*

- *A pocket-size automatic language translation system is used in practice for conversing without any knowledge of the language of the other person.*

- *Science museums are widespread in Germany in which the skills necessary for scientific tasks are promoted with the aid of natural history and scientific educational methods in a play environment.*

- *A refresher and training system for the occupational development planning of middle-aged and older people is universally established, in the framework of which they can acquire new specialist knowledge and technical qualifications.*

The following topics have decreased relatively in importance:

- *A method of carbon dioxide immobilisation is in practical use which reduces CO_2 emissions by half.*

- *An automatic sorting technique is in universal use which separates normal household refuse by degree of hardness, relative density, moisture and colours into combustible substances, metal and glass.*

- *A navigational guidance system is set up on such a comprehensive basis that in bays with busy traffic, such as the German Bay, all ships are subjected to a full check, thus guaranteeing safety and trouble-free traffic.*

- *A process which dissociates water by means of solar radiation is in practical use.*

- *Irrespective of the type of material, the hetero-epitaxy technique is developed on silicon crystal wafers.*

This list indicates that environmental and resource problems today no longer have the same high status compared with other subjects as they did approximately five years ago.

One last question which comes directly to mind and which can also be answered by virtue of comparably-phrased questions, focuses on changes in the competitiveness of Germany in research and development.

In relation to the parent population of 113 topics asked in identical fashion in both Delphi studies, Germany has made the furthest strides forward in the development of *science museums in which the skills necessary for scientific tasks are promoted with the aid of natural history and scientific educational methods in a play environment.* The catching-up process in the case of *remote monitoring and control systems to guarantee greater security of supply lines* is assessed in a similarly positive manner. The position of Germany's research and development is also seen as being markedly more optimistic in relation to *automatic translation devices, systems for automatically summarising documents*, deep-drilling technology *from the sea floor*, the clarification of the *pathogenesis of Alzheimer's disease, flexible floor plan* solutions in the residential sector and *refresher and training systems for occupational development planning*.

Setbacks are to be feared for Germany's research and development in space technology in relation to the development of *a system for verifying gravitational waves* and the construction of *solar power plants with extra-large solar cell panels*. However, Germany is also losing ground in sectors of materials research, e. g. in *polymers which have memory, recording and switching capabilities*, in *highly-magnetic materials with saturation magnetisations of more than 3T and in relatively long fibre optic materials for transmission rates of 100 Gb/s*.

As the Japanese colleagues have the longest experience in foresight studies, it is interesting to trace back some of their topics in more detail. One example is more or less a failure of prognosis: earthquake prediction is not possible, and its realisation is always prolonged into the future (see Cuhls 1998). The second example is photovoltaics. Other examples like the fast breeder and the fax machine can be followed elsewhere (Cuhls 1998). Earth-quake prediction is a typical Japanese example for forecasting (in the sense of prognosis), as Japan is located in an area where many earthquakes occur and people must be evacuated in case of emergencies. This is therefore a topic of general interest – not only for companies but also for politics, or the society in general. But earthquake warning is difficult. And if earthquakes can be predicted – how early can persons be evacuated and damages be prevented? How can the responsible institutions be convinced to believe in that warning so that the right measures can be taken?

The following questions have to be answered for an earthquake prediction and warning (Hurtig/Stiller 1984).

- Where does the earthquake take place (epicentre) and how deep is it under the surface?
- When does it take place? One distinguishes between short-, medium- and long-term forecasts.
- How strong is it (magnitude)?
- What will the intensity be at a certain place in correlation to the distance from the centre of the earthquake and the local surface quality?
- What follow-up problems (phenomena) can occur (like tsunamis, landslides etc.)?

Earthquake prognoses are based on quantitative models using seismic data, but can also take the unusual behaviour of animals into consideration or weather phenomena that occur before the earth is trembling. The last mentioned methods are more the Chinese approaches and have to struggle with the connotation of not being "scientific" enough. Until now, no earthquake has been predicted exactly by location, time and magnitude.

Thus, in all Delphi studies, earthquake prediction is a topic, but not exactly with the same wording. In the first Japanese Delphi study of 1971, one earthquake topic (no. 1/74, see fig. 6-1) is found in the field "society", and can be translated as follows: *A technique is established to forecast if in the area of prefectures (Fu or Ken) an earthquake (magnitude 6 or higher) occurs within one month (or not)*. The second topic is found in the field "information" concerning the needs of economy and society: *A forecasting technique is established, with which an earthquake of a magnitude of 6 or higher occurring within the next month can be predicted in advance* (no. 2/3, see fig. 6-1). The formulations are not identical but the issue as such is the same.

But although the importance of both topics was estimated as very high (Index about 91 and 93 on a scale up to 100), the time of realisation was estimated differently. The experts in the field "society" were not very sure about their estimations so that they came up with a range from 1983 up to a time after 2000 (the horizon of the first Japanese Delphi survey) with a median of 1996. For the second topic, the estimation ranged from 1986 and shortly after 2000 with a median of 1994 for the first survey round.[9] In the last round, these discrepancies were even larger (see fig. 6-1): the first topic was regarded as realistic between 1988 and a long time after 2000 (median 1997), and the second one between 1988 and 2000 with an earlier median of 1994.

This can – of course – be a statistical artefact as in the field "society" only 113 persons answered, but in the field "information" 400 persons judged on the matter. It is interesting that in the first case the persons with a very high expertise are more optimistic (1981-1987-1997), but in the second case estimated in the range of the others or later (1988-1992-after 2000). This discrepancy cannot be explained, because the respondents' knowledge base can no longer be evaluated, as their names and functions are not known anymore.

---

9   The first Japanese Delphi study had three so-called rounds: the first round was the preparation round for the topics, the second one was the "first survey round with questionnaires", and the third one the "second questionnaire round" with the end results. Starting from the second Japanese Delphi study, the surveys only had two questionnaire rounds although the topics were prepared and re-arranged in expert committees, too.

Figure 6-1:   The estimations on earthquake prediction in the different Japanese Delphi studies

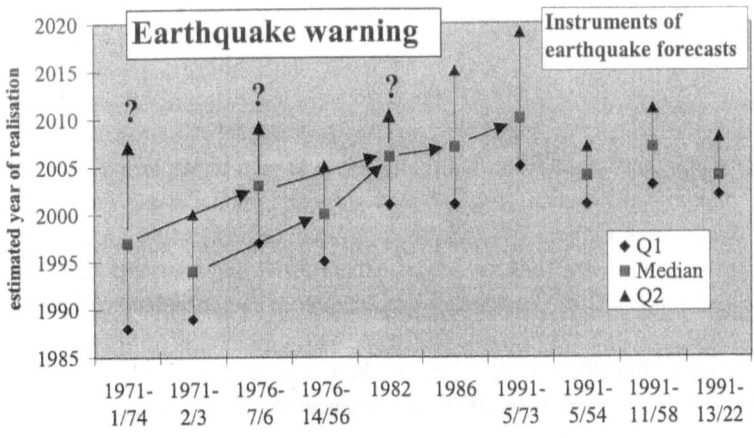

Delphi survey (year-number of topic)

The probability that the topics can be realised at all is estimated with 74:26 and 78:22 (this adds up to 100 % each), which is very similar in both cases. Many experts (26 %/ 29 %) assessed earthquake prediction as technically difficult, other problems are rather not mentioned. Among the measures for the state government, funds for investment (27 %/ 37 %) and the research and development system (34 %/ 28 %) were mentioned, whereas 22 or 24 % of the experts said, special support is not necessary, and 22 or 18 % called for better education and human resources in the field.

In the second Delphi survey of 1976, the same issue was asked for with the topic no. 1/74 (same wording), but again: in the field "security" and the field "information". Again, the experts were not sure when earthquake prediction would be possible. In the last round, the 53 "security" experts estimated a time between 1997 and after 2005 (median 2003), and the 50 "information" experts between 1995 and 2005 (median 2000). This is in both cases later than in the first Delphi rounds (1993 to after 2005, and 1990 to 2005) as well as later than in the previous 1971 Delphi study. But the importance rate remained extremely high (index 91 and 94 with no one saying "low" or "no importance"). Again, the *technical* problems were regarded as the obstacles. R&D was defined as a task of the state government, and more experts voted for own research in Japan (80 %/ 57 %) than for international cooperation. In more detail, funding (30 %/ 70 %) and an improvement of the R&D system (55 %/ 12 %) were asked for.

In the third Delphi study of 1982, the time of realisation of the same topic was only a little bit "postponed" into the future. But experts were again unsure. They estimated the time between 1996 and long after 2010 (median 2006) in the first round, but postponed it with a greater consensus to 2001 and after 2010 (median again 2006) in the second. The importance index was about 95 (extremely high). New in the third survey were instruments for earthquake prediction like *sensors to measure temperature, pressure, gas, and humidity* and *satellites*.

In the fourth Delphi survey of 1986, the earthquake topic was discussed in the field "earth and marine sciences" and reformulated to a certain extent because of comments concerning magnitudes in the third Delphi to: *a forecasting technique is developed, that allows to announce earthquakes of a magnitude of more than 7 on the Richter scale some days in advance.* This means the forecasting precision of the time must be better (some days in advance), but for stronger earthquakes. And the location was not asked for anymore, which seemed to be redundant to the experts. But the experts were not optimistic, anymore, and estimated a realisation between 1999 and long after 2015 in the first, and between 2002 and 2015 in the second round (median in both rounds 2007). Importance and measures to be taken were judged similarly as before.

In the fifth Japanese (NISTEP 1992) and the first German Delphi study (BMBF 1993), the same topic as in the previous Delphi was asked for – and again postponed into the future: between 2005 and 2019 (median 2010) in Japan. The importance was rated as very high, the technical obstacles were mentioned (81 %) as well as problems of funding (47 %), but international co-operation was rather denied (index 66). One reason might be that 48 % of the respondents estimated Japan as the strongest country in the field (with the strongest competitor USA only 38 %). The German experts were very optimistic with a time estimation of 2002 to 2013 (2005) with a much greater variety in the first round and relatively high demand for international co-operation. But the small sample of 19 respondents was not representative, especially because only three of them were experts working in the field. This shows that in Germany earthquake prediction is not one of the priorities.

In the fifth Delphi, topics dealt with instruments for earthquake prediction which were also rated as very important (see fig. 6-1), e. g. no. 5/54 from "earth and marine sciences": *The nation-wide installation of bore-hole observation equipment integrating various types of gauges (e. g., seismometers, tiltmeters, and strain-gauges) can be used for earthquake forecasting.* (possible realisation: 2001-2004-2007), no. 11/58 from "urbanisation and construction": *Nation-wide networks for detecting earthquakes, enabling dissemination of disaster preventing systems, which transmit information on earthquakes at distances of about 50 km or more in advance, are established.* (possible realisation: 2003-2007-2012), or no. 13/22 from "space": *an accuracy down to 1 cm or less in measurement of crustal movements based on very long baseline inter-ferometres (VLBI), satellite lasers, and inverse laser ranging is realised, improving the precision of earthquake forecasting* (possible realisation 2002-2004-2008).

In the latest, the sixth Japanese Delphi (NISTEP 1997), the reality had shown that earthquake prediction and all "earthquake forecasting plans" (JETRO 1991) did not function, although the instruments for measurement (see above, seismometers, GPS, sensors, information systems) had meanwhile improved drastically. In Japan, scientific earthquake prediction is based on a very dense network of measuring seismic data. The Japanese Science and Technology (STA) is responsible for earthquake research with its National Research Institute for Earth Science and Disaster Prevention (NIED). Also the Meteorological Agency (belonging to the Ministry of Transport) and its Meteorological Research Institute are devoted to seismic research as well as the Earthquake Research Institute (ERI) of the University of Tokyo and the Geographical Survey Institute (GSI).

Nevertheless, the Great Hanshin Earthquake, which nearly destroyed Kôbe and killed more than 6,000 persons, found the western region of Japan (Kansai) unprepared. The earthquake was not expected to occur at this location, there was no warning at all, all programmes to help failed – and in the end, the increase of the budgets for earthquake disaster prevention was highly criticised (Cuhls 1998, p. 307-311, Swinbanks 1994, p. 9, and 1995, p. 373).

The sixth Delphi study in Japan was conducted shortly after this large earthquake. That the experts were still under this impression is reflected by their new estimations. The realisation for the topic concerning general earthquake forecasting (*technology capable of forecasting the occurrence of major earthquakes, with a magnitude of 7 or above, several days in advance is developed*) is shifted to the longer-term future (2016-2023-long after 2025) but assessed with a very high importance (index 92 on a scale of maximum 100) with "high effects on peoples' needs" (85 %). This time, it was also asked for the *practical use of a mid-term (5-10 years in advance) prediction technique for large-scale (magnitude 8 or stronger) earthquakes based on analyses of the distribution of strains in the earth's crust and past earthquake records*, which was regarded as possible between 2012 and 2023. But this was not the only effect of the real earthquake occurrence. Beneath the topics concerning the instruments for earthquake prediction, which were again included in the study, it was also asked for measures to take if an earthquake occurs, e. g. the following (NISTEP 1997).

- *Disaster forecasting and information transfer systems are developed incorporating studies in social and behavioural psychology, in order to prevent panic in big cities in event of major earthquakes or fires.* (no. 3, Urbanisation and Construction)

- *A nation-wide network for detecting earthquakes is developed and a disaster prevention system is in widespread use in Japan that gives advance warning of earthquakes at a distance of at least 50 km.* (no. 4, Urbanisation and Construction)

- *Online data bases on natural disasters necessary for risk management are in practical use in Japan.* (no. 7, Urbanisation and Construction)

- *Emergency response systems to deal with disasters (e. g. fires and earthquakes) occurring in deep underground facilities are in use in Japan.* (no. 9, Urbanisation and Construction)

- *Technology is in practical use in Japan to effectively control and absorb vibrations in massive structures caused by winds and earthquakes.* (no. 16, Urbanisation and Construction)

- *A system is developed that detects the initial mild tremors of an earthquake at appropriate locations, and safely stops trains as necessary to avoid places that have a high risk of collapse (because of the earthquake).* (no. 7, Transportation)

- *Integrated building management systems and home security systems are in practical use which are linked to an earthquake detection system and take the necessary safety measures to protect human lives in the event of a non-direct-hit earthquake, taking advantage of the time lag to the arrival of seismic waves.* (no. 63, Communication)

This shows the disappointment regarding the prediction capability but also Japanese pragmatism. As the earthquake cannot be predicted or "avoided", buildings, facilities, materials or systems have to adapt to the situation. That means, the struggle for new technology shifts to the closer, easier to realise technological solutions of the problem. The Kôbe earthquake was a great lesson in this regard, especially in the field of house building (the new buildings constructed according to the high standards were damaged, but did not fall down, the older houses collapsed, and some cases of corruption turned out to be the cause of other damages, e. g. when cheaper, low standard materials were used or the construction was not according to the new standards).

Although it must be acknowledged that the possibility to make earthquake predictions has shifted into the future and therefore, all forecasts seem to be a failure, the Delphi results have a logic in themselves. The different instruments for quantitative measurements are seen as realisable – and have already been realised to a certain extent. They seem to be the technical presumptions. At first, the networks with their equipment have to function, and then, the models and simulation can work. One problem of the whole system is still the theoretical models that could be used, another problem is the structure of the different institutions in Japan, which make the transfer and gathering of information difficult (Cuhls 1998, p. 311).

For foresight studies, the lesson to be learned is to ask for clear technologies and their application, e. g. instead of asking for "earthquake prediction technology" (in general) to ask more concretely for e. g. the "nation-wide network of seismometers that gives advance warning of earthquakes". Another lesson is that there are topics that cannot be *predicted* at all – in spite of all attempts and in spite of remaining problems that ask for a solution. An indicator can be that the experts in a Delphi study claim to be very "unsure" (precision, broad time frame Q_2-Q_1) and that their answers are very diffuse on the time-scale.

But why is this type of erroneous forecast nevertheless helpful for technology policy? Even before the Kôbe disaster, i. e. without a strong stimulus for R&D in earthquake detection, the foresight activities kept this issue on the agenda in science and technology, admittedly on a low level. Every five years, an earthquake prognosis plan was elaborated, and MITI financed geological surveys. In 1993, before the Kôbe earthquake, a special measures law for earthquake disaster prevention passed the Diet. The R&D budget was already at about 16 billion Yen in 1996 (STA 1997). Foresight studies and their ever shifting of realisation times were helpful in pointing at the unresolved issues in years of no earthquakes. They also pointed to the necessity of being prepared for an earthquake although after many years without earthquakes the attention gets lost. Delphi surveys in these cases have the objective to remind of the unsolved problems in the field. Until then, there are no other possibilities than to build as safely as possible, to have insurance and to prepare for something, even if one does not know if it will ever happen.

This chapter demonstrates that in many cases, the time frame of innovations are prolonged into the future. That means, an update of the data is necessary from time to time. Five years (like in the Japanese cases) seems to be a good time span for this update. Thus, single actions are not as effective as continuous foresighting.

7   German-Japanese Comparison

As already described, 323, i. e. just a third of the topics were equivalent to the items of the sixth Japanese Delphi Report. This makes possible comparisons in the assessment of future technology. Is the assessment identical to the innovations? Or do we have "national blinkers" on? In terms of ranking, leadership in research and development is essentially the same in the Japanese and German assessment, but on the other hand, the experts of both countries each consider themselves further advanced than their colleagues in the other country.

In Delphi '98 in fact, it is no longer possible to compare all categories with Japan (as it was for Delphi'93). But to compare the time horizon is quite fruitful. Other categories that can be compared are:

- The *expertise* in the field. Here, exactly the same criteria are asked for.

- The *importance for the economy* and *for the enlargement of human knowledge*. The other criteria in this category are not equivalent but similar, and of the same number. As multiple choices were also possible, we can compare if the percentages of the criteria are similar or differ to a great extent. It is impossible to compare the percentages directly, but to rate high versus low numbers.

- *Measures* like *better training or education* and *change in regulations*. The measures are also quite similar but the wording is understood differently in the two countries. But as "training" and "regulation" are also exactly at the same position in the questionnaire, as the number of criteria is the same for "measures", and as multiple answers were possible, the percentages can be compared directly.

- *Follow-up problems* are asked in a very similar way in Japan and in Germany, so that the percentages can be compared directly. There is only a difference in wording for *social or socio-cultural problems* in the German, and *ethical problems* in the Japanese questionnaire, which means that this single criterion cannot be compared directly as the ethical dimension is included only as a part in the German definition.

We start here with a comparison of the assessments on the time of realisation concerning all comparable topics, then we will pick out some differences in single topics and fields. In some cases, the explanation for differences is obvious. In others, no easy explanation can be given.

7.1   All topics

In 86 % of cases, i. e. for 277 of the total of 323 comparable topics, the assumed period is very similar. The median differs here by less than five years. (Reminder: only five-year periods could be chosen.) This shows how similarly the future of science and technology is viewed at an international level. If one looks at the time distributions onto the different five-year intervals (fig. 7.1-1), they differ only marginally in the comparison between Germany and Japan. (In general, the German experts are very slightly more optimistic). This shows that at least the chosen identical topics, of which a particularly large number originate from the information and communications technology sector (see table 7.1-1 and chapter 7.2), are judged similarly at an international level. The experts expect a particularly large number of realisations in the medium-term future, but particularly few in the very near and the distant future.

Figure 7.1-1:   How are the assessments of the time of realisation distributed in the German-Japanese comparison?

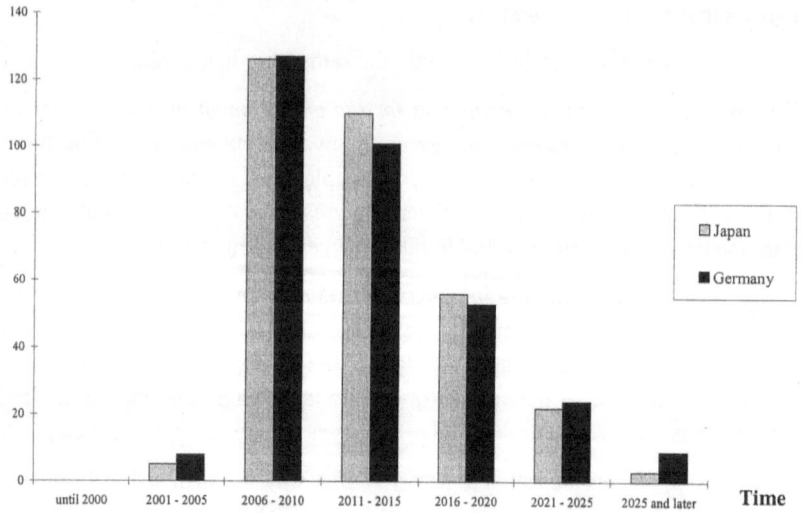

More interesting, however, are the 45 topics (14 %) with a very varied assessment. The evidence shows that one half of these (23) is judged more optimistically by the Japanese, the other (22) is judged more optimistically by the German experts. No structural or methodologically-determined differences in the judgement can therefore be assumed, only differences with regard to content.

What is striking is that the topics listed in table 7.1-1 with the greatest differences in terms of time assessment are in any event considered possible in 10 years at the earliest. And the further in the future, the more difficult the assessment is and the more variable it therefore turns out. However, there are also individual explanations for the differences in terms of content, e. g. the heavy investment of the Japanese in marine research and, resulting from this, a greater optimism for being able to *cultivate microorganisms in the sea* or have *wave power stations for energy recovery.*

Table 7.1-1:   Topics showing the greatest differences between the German and Japanese evaluations on the time of realisation (from all fields)

Field (Germany)	Topics	Median Japan	Median Germany	Differ- ence
Environment	The cultivation of microorganisms in the sea and on the sea bottom is in widespread use for technical applications.	2010	2022	12
Energy	Wave power stations to utilise the energy of wave movements are in practical application.	2009	2019	10
Management	Space factories are built to produce new materials, utilising high-vacuum and weightlessness.	2017	after 2025	>9
Agriculture	Genetically modified plants and microorganisms resistent to NO_x or which reduce the NO_x content in the air are applied in environmental protection.	2012	2021	9
Information	After the elucidation of the physiological and psychological mechanisms causing human defects, an early warning system is practically applied.	2015	2024	9
Big Science	High-efficiency geometrical optics for analyses of radiating equipment, e. g. substances under the earth, using electron or positron storage rings with a radiation emittance under 1 m pGy are in use.	2024	2015	9

German experts think that *factories in space* are only possible at a later date (after the time horizon of this study, in 30 years) or not possible at all (11.3 % of respondents say this); the USA is the solitary front-runner in this research. People also think similarly in Japan, but otherwise they are more optimistic. Only 5.4 % of the experts think space factories are never achievable; the others assume the first practical examples will be seen between 2012 and 2022. However, they believe better training of personnel, new institutions and facilities as well as more financial support is needed in the sector. According to the Japanese comments, these projects are very important for the economy and scientific progress.

A greater degree of scepticism towards the *use of genetically modified plants and microorganisms* surely results from the rather chilly German attitude to genetic engineering in general which has prevailed to date. Consequently, even the experts do not believe in an early realisation. The German experts are likewise sceptical with regard to the question whether an *early warning system can exist which warns people of defects in good time.* It is not only the necessary clarification of different physiological and psychological connections which appears difficult in this instance, but their "technical" use is not viewed as euphorically as in Japan.

Being able to use high-efficiency geometrical optics for analyses of radiating equipment with certain electron or positron storage rings, is something that the German researchers in turn think they will live to see earlier than their Japanese counterparts, even if it is only in 15 to 20 years' time. In keeping with this, the German experts consider their own state of research (along with the US American) to be particularly far advanced. Without an appropriate research infrastructure and international co-operation, however, people think that they will be unable to solve this problem even in Germany. This is the only case, where the Japanese are definitely more pessimistic in the time frame than the German colleagues.

Selectively, there are differences in the time estimations, interestingly also in the field of Information and Communication, which is very international in nature. But in general, the differences in the time of realisation are rather marginal.

Other interesting findings concerning all topics are the following:

There are fields in the German Delphi '98 in which the expertise is rather low (see above), like Mobility and Transport, Construction and Living as well as Services and Consumption. Interestingly, this low expertise is rather similar in Japan, especially for those single topics where it is low in Germany. This can have different reasons: maybe there is only selective expertise available and the problem is the same in Japan and Germany in this field. Or like in Services and Consumption, it is generally difficult to identify experts who have a certain knowledge of the items. In this book, this cannot be checked in detail.

Another point is that in most fields in Japan *better education* is called for. Whereas in Germany, this measure is called for in selective cases, in Japan, this seems to concern whole fields and is therefore generally higher: Chemistry and Materials, Space, Energy and Resources, Environment and Nature, Management and Production, Agriculture and Food, Health and Life Processes, and Information and Communication. Especially the last field shows that not only Germany is confronted with a "green card" debate on the question whether they should invite engineers and specialist personnel from other countries. The question if persons should be let into the country is widely discussed in Japan, too.

As better education is called for so often, it clearly demonstrates that the Japanese patriarchical structures in the universities and other research facilities have to be improved. First attempts have been made with the re-structuring of the political administration (e.g. fusion of the Science and Technology Agency and the Ministry of Education) as well as the evaluation activities (Kagaku Gijutsuchô 1998; Tsuchiya 2000) and plans for re-structuring the Japanese public universities (Hirowatari 2000).

In general, there are more *follow-up problems* mentioned in Germany, especially the problems for the environment are regarded as severe. Even in fields like Management and Production, the Germans expect environmental problems. This cannot be explained simply and directly but would have to be checked in every single case. But let us look at the single fields in more detail.

7.2   Information and Communication

For most topics in the field, the Japanese experts claim to have a slightly higher *expertise* than the Germans, but as this is a self-estimation it is difficult to judge the reason and to find out if this slight difference really has an impact on the result.

The importance criteria are assessed similarly in Japan and in Germany: both expert groups assess the relevance *for the economy* as especially high. Table 7.2-1 shows some of the topics with greater differences. Note that the percentages cannot be directly compared. It can only be stated if the importance is rated significantly higher in one country.

On the other hand, the Delphi topics will only have a small impact on the *enlargement of human knowledge*. This concerns obviously topics which are related to applications whereas the more basic questions like the *bio computer on the basis of a new algorithm*, the *clarification of the mechanisms in the creative capabilities of human beings so that they can be used in information sciences*, and that *computers can read the information that is stored in the brain by using electrical and electro-mechanical techniques* are rated to have a higher impact on human knowledge.

Table 7.2-1: Topics with differences in the estimation on the importance for the economy (in percent, for the calculation see chapter 2.5)

Topic	Germany (%)	Japan (%)
Optical wireless communication using indirect and dispersed light in closed rooms, by which PCs and monitors can be joined in networks without wires is in widespread use.	91	63
Due to the development in multimedia communication using the Internet and Intranet (e-mail, WWW, telephone conferencing etc.) general office work at home, with the exception of meetings and negotiations, is very common.	90	72
A process for high-quality speech synthesis is in practical application which converts written information automatically into speech with almost the quality of the human voice.	78	59
Systems will be developed which can actively recognise and explain contexts (metaphors, synonyms, objections, ambiguities etc.) in information.	60	41
Techniques will be developed with which one person can be identified out of more than 10,000 by means of picture information from video film.	46	77
A portable communication device that can send and receive alarm signals in emergencies is in general use.	41	26
A computer network is in wide application by which a virtual reality (e.g. a virtual mall) is suggested, allowing any number of people, although spatially dispersed, to share a common perception simultaneously.	52	82

Although there are only few topics with a difference in the estimated time of realisation, there are some with a difference in the median that is higher than four years (table 7.2-2), especially in the field of Information and Communication. The gray parts in the following tables mark the differences or stress the specialities found. In the case with the greatest difference (see the first three topics) the German experts are more pessimistic, in all other cases with high differences, the Japanese experts estimate a later realisation time.

Table 7.2-2: Topics in Information and Communication with a difference in the time of realisation (median) of five or more years (grey boxes stress the difference)

Topics	Germany time				Japan time			
	Q1	Median	Q2	never (%)	Q1	Median	Q2	never (%)
After the elucidation of the physiological and psychological mechanisms causing human defects, an early warning system is practically applied.	2017	2024	2028	41	2011	2015	2019	1
A writing recognition technology is in common use, with which over 99% of handwritten Chinese characters can be read (an example of a particularly difficult type of pattern recognition).	2012	2016	2020	1,6	2005	2008	2011	7,7
Facilities are developed which are capabale of automatically compiling summaries or excerpts of books or documents (in any length acc. to need).	2011	2016	2022	16	2006	2009	2014	7,1
Systems will be developed which can actively recognise and explain contexts (metaphors, synonyms, objections, ambiguities etc.) in information.	2015	2020	after 2025	17	2009	2015	2023	16
Components which integrate sensors, controllers and actuators find practical application in micro-machine technology.	2002	2005	2008	0	2009	2012	2017	1,5
Fixed and overwritable RAM of more than 100 Gb are used in practice.	2007	2010	2014	0	2013	2017	2022	4,1
Opto-electronic integrated circuits (OEIC) are widespread, in which several optical elements are integrated with beam waveguides on semiconductor chips.	2004	2007	2010	0	2009	2013	2017	3,7
LSI circuits with response times of under 1 ps are in practical use.	2006	2009	2013	2,1	2012	2015	2020	7,2
Surface shining laser arrays 1,000 x 1,000 pixel are in practical use for e.g. optical coupling.	2005	2008	2011	0	2010	2014	2019	4,7
Optical switchboards are in use which have a ca-pacity of 10,000 connections for video terminals.	2007	2009	2013	0	2012	2015	2019	2,7
High-frequency broadband amplifiers in semi-conductor technology are practically applied in the frequency range over 1,000 GHz .	2007	2009	2013	2,7	2012	2015	2019	1,8
Biosensors which can identify a single molecule are in practical use.	2009	2013	2017	9,6	2014	2019	2024	3,2
A manufacturing technology is practically applied with which - adapted to the necessary specifi-cations of the system level - high-performance LSIs of several 100 K-gates are designed completely automatically.	2007	2009	2012	2,4	2010	2014	2019	1,7
Optical multiplexers are used with which 200 canals of 100 Gb/s can be sent over a beam waveguide in multiplex mode.	2007	2009	2012	1	2011	2014	2019	3,7

Both the Japanese and German experts regard the USA as the leading country in the world. The Japanese experts rate the USA to be better in R&D than the German experts think they are. No direct comparison can be made for Germany and/ or the European Union as the Japanese only asked for the European Union in general. Additionally, they assessed the former Soviet Union as not being a major player in the field.

As already mentioned for measures, the Japanese experts call for *better training and education in* nearly all the themes and mention it much more often than the German experts. The only exceptions are *due to the development in multimedia communication using the Internet and Intranet (e-mail, WWW, telephone conferencing etc.) general office work at home, with the exception of meetings and negotiations, is common,* for which 53 % of the German experts and only 15 % of the Japanese think better education is necessary, as well as *remote teaching systems will be in general use which makes possible the training and continuing education of the population from home,* with 58 % of the German and 19 % of the Japanese experts calling for an improvement.

The last mentioned example is also one for the difference regarding *regulation*. The Japanese assess more often (54 %) that they might need changes in regulation than the German experts (only 20 %). Regulation changes are asked for more often in Japan than in Germany when whole systems or networks are concerned (card systems, electronic trade, mobile communication). In Germany, more regulation is mentioned for *systems of virtual examinations in public networks* (63 %; Japan: 46 %).

The German experts see many cultural and social follow-up problems (not comparable to the Japanese criteria of ethical problems which are mentioned less frequently), but selectively also environmental problems, e.g. *temperature-resistant logic ICs for use at temperatures up to 500°C are in practical application; high-temperature superconducting materials are in general use in passive circuits of information systems which work in the dwarf waves area; low voltage PCs are in widespread use which can be operated for one year on a single round cell battery; biosensors, which can identify a single molecule, are in practical application; a system is widespread which gathers different supra-regional information telematically for weather prediction, environmental monitoring, traffic surveillance etc. via satellites and transmits them to earth.*

Information and Communication is a field in which many (63) topics can be compared, and relatively many, but very selective differences could be detected. Most of them need more research in order to verify if there is a real difference or if the selection of experts influences the judgement. Therefore, no straightforward explanations can be given.

7.3   Services and Consumption

The field of Services and Consumption is so broad that it is more difficult than in other fields to identify "experts" with sufficient "expertise" to be invited to participate in the Delphi surveys. Therefore, the expertise both in Japan and in Germany is relatively low for the 20 topics that can be compared[10]. A major difference can be observed for the topics *systems to purchase multimedia information (also scientific information) on demand are widespread and available worldwide in decentral networks; electronic newspapers and electronic museums are widespread, which can be compiled according to individual interests; and societal rules on multimedia authors' rights will be agreed on so that the production and circulation of multimedia information increases,* for which the Japanese experts claim a higher expertise.

Concerning the importance for the economy, the Japanese experts assess *portable devices are in common use with which newspapers can be transmitted to information kiosks and so bought without physical contact; electronic elections will be carried out from home; an "electronic parliament" (electronic state assembly) operates and sits in parliamentary television programmes, so that decisions on draft bills (regulations) are made with the aid of electronic plebiscites; electronic newspapers and electronic museums are widespread, which can be compiled according to individual interests* and *for the first time robots are the sparring partners of people, e.g. in martial arts* (but on a lower scale) as much more important than the German experts, whereas the German experts think that *robots which guide blind people e.g. in railway stations or shopping centres will be in practical use; an "intelligent" wheelchair will be used in practice which automatically adapts to stairs, escalators and rising ground* (these first two topics are generally not very important for the economy) and *due to progress in forecasting technology for landslides as well as mud and rock slips, the number of deaths will be considerably reduced worldwide* are more important for the economy (and mainly society).

For the time of realisation, the German experts are sometimes more pessimistic than the Japanese (table 7.3-1).

For most of the topics, both German and Japanese experts think the USA are the most advanced in R&D. But for the topics concerning the *sports robot, the robots that can save people in case of a disaster,* and *the robot that guides blind persons,* the German experts think that the Japanese experts are the leading in the world, even better than the Americans. Interestingly, the Japanese do not claim that position for themselves: for the *sports* and the *disaster robot,* they think the Americans

---

10   Note that there was no explicit field called Services and Consumption in Japan. The topics that can be compared stem mainly from the fields Information, Communication and Electronics.

are in front, for the *robot guide*, they think the experts from the EU (maybe the Germans?) are the most advanced[11]. This seems to reflect mutual prejudices.

Table 7.3-1: Topics in Services and Consumption with a difference in the time of realisation (median) of five or more years (grey boxes stress the difference)

Topics	Germany time				Japan time			
	Q1	Median	Q2	never (%)	Q1	Median	Q2	never (%)
Robots which guide blind persons e.g. in railway stations or shopping centres are in practical use.	2009	2013	2018	8,1	2004	2006	2010	0
Science museums are widely found in Germany/ in Japan, in which the abilities for scientific work are encouraged in play, using natural history and scientific training methods.	2009	2013	2017	10	2003	2007	2010	1,5
Electronic elections can be carried out from home.	2009	2014	2019	18	2007	2009	2013	6,8
An "electronic parliament" (electronic state assembly) operates and sits in parliamentary television programmes, so that decisions on draft bills (regulations) are made with the aid of electronic plebiscites.	2014	2019	2025	47	2010	2013	2018	21,3
Due to progress in forecasting technology for landslides as well as mud and rock slips, the number of deaths will be considerably reduced worldwide.	2008	2011	2015	11	2012	2016	2021	5,2

Selectively, for some topics regulation changes are claimed. Not for every topic, but for single ones, the Japanese regard better education as necessary, whereas here, the Germans do not rate the importance of education that high (table 7.3-2).

In this field, the German experts expect many social and cultural follow-up problems, but the criteria cannot be compared. Greater security problems are feared for *digital money* and the *intelligent wheelchair*. In the first case, the Germans, in the second, the Japanese are more pessimistic. Environmental problems are expected

---

11 The Japanese experts often admitted to having problems in judging the European Union: Should they judge the whole EU or single countries in the EU? Sometimes they just do not know which country belongs to the EU, especially as there were so many changes in the last years. In the 7th Japanese Delphi, which is currently being conducted, this problem is even worse.

only by the German, not by the Japanese experts for the topics concerning *weather forecasts and the forecasts of landslides, mud and rock slips*. The German assessment cannot be explained. Maybe the experts fear that countermeasures based on weather forecasts evoke environmnetal problems. Another possibility is that they just misunderstood this criterion, but in this case, it is doubtful that such a misunderstanding occurs only once.

Table 7.3-2: Japanese find more often that better education is necessary[12] (in per cent of all measures mentioned, for the calculation see chapter 2.5)

Topics	Germany Measures Better Education (%)	Japan Measures Better Education (%)
Robots will be developed which can recognise, seek out and help people in the case of disasters.	12.9	55.3
Due to progress in forecasting technology for landslides as well as mud and rock slips, the number of deaths worldwide will be considerably reduced.	17.2	54.3
Weather predictions of up to one week in advance will be possible with more than 95 % accuracy.	13.8	53.5
Security and surveillance systems are implemented with the help of robots which possess general information and functions to prevent crime.	7.8	39.7
Electronic newspapers and electronic museums are widespread which can be compiled according to individual interests.	16.2	38.4
Portable devices are in common use with which newspapers can be transmitted to information kiosks and so bought without physical contact.	18.6	32.5
For the first time robots can be used as sparring partners of human beings, e.g. in martial arts.	4.4	26.5
Appliances are widespread in normal (German/ Japanese) families by means of which travel, films, sport etc. can be virtually experienced.	9.4	25.2

---

12 It has to be taken into account that the rates for the measures are in general lower than those for other criteria, so that rates around 50 % can be considered relatively high.

7.4　Management and Production

The 13 topics of this field that can be compared directly are mainly concerned with technical matters whereas in the German questionnaire most of the topics concerned organisation or management matters. Therefore, in the technical aspects, the expertise was not very high in Germany or in Japan.

Most of the topics mentioned are extremely important for the economy of both countries. Differences can be observed for the topic *manufacturers of durable consumer goods are obliged by law to take back and dispose of their products at the end of their operating life, whereby a recycling system with planning, production, collecting and recycling or re-use is in common use, by means of which an almost complete material closed loop emerges* which is regarded as important by 65 % of the German and only 40 % of the Japanese experts. The German result corresponds to a certain extent to the bi-annual survey "Innovations in Manufacturing" made by ISI (Dreher/ Schirrmeister 2000).

Another difference can be observed for the topic *programmes are in widespread use, in which the learning ability of living organisms is simulated and thus the performance of these programme can be automatically increased* for which the Japanese experts estimate the importance for the economy higher than the Germans, but on the contrary, the German experts rate the importance for the enlargement of human knowledge higher than their Japanese colleagues. The only topic from this sample in which the importance of human knowledge is assessed to be high in general and also higher in Germany than in Japan is *an interactive system with artificial intelligence (AI) is widely in use which gives impulses to developing ideas or points out correlations.*

The time expectation shows differences for three topics (table 7.4-1). The first one concerns again the *recycling and re-use system*. The second topic is the *factory in outer space* with a very long-term time horizon, for which the German experts are just more pessimistic than the Japanese. The third gap in time is not that huge but concerns a short-term development of e-commerce with a great importance for the economy. One reason for the difference can be regarded in the present use of order systems which are more widespread in Germany, where shops are closed on Sundays, than in Japan in general so that the ordinary way of ordering is just replaced by an order via internet with the special feature of being able to design an individual product.

Table 7.4-1:   Three topics with a difference in the estimated time horizon (grey boxes stress the difference of the medians)

Management & Production	Germany				Japan			
	time of realisation				time of realisation			
	Q1	Median	Q2	never (in %)	Q1	Median	Q2	never (in %)
Manufacturers of durable consumer goods are obliged by law to take back and dispose of their products at the end of their operating life, whereby a recycling system with planning, production, collecting and recycling or reuse is in common use by means of which an almost complete material closed loop is achieved.	2002	2005	2009	1,8	2009	2013	2017	3,6
Space factories are built to produce new materials, utilising high vacuum and weightlessness.	2020	after 2025	after 2025	11,4	2013	2017	2022	5,4
Order systems can be used from the home which enable the consumer to design products individually (e.g. a car to own specifications).	2003	2005	2009	0,6	2007	2010	2016	4,7

Although the highest R&D level is not directly comparable, it can be stated that both countries rank the USA first. Interestingly, the Japanese experts rate the R&D level in their own country significantly higher than the German assessment does (table 7.4-2).

Like in many other fields, the Japanese experts call for better education much more often than the German experts. The only exemption in this sample is the e-commerce *order system for individual products*. There seems to be a larger education deficit in Germany in this point. Even in this field, the German experts expect more environmental problems than their Japanese colleagues. This seems to be a general attitude.

Table 7.4-2:   Differences in the R&D level (in per cent, for calculation see chapter 2.5, note that the countries are only indirectly comparable, grey boxes stress the difference)

Management & Production	Germany highest R&D level					Japan highest R&D level				
	USA	Japan	Germany	other EU-country	other country	USA	EU	former USSR	Japan	others
Polymers which possess memory and recording abilities, logical elements (switching) are utilized in electronics and informatics.	90	49	31	3	1	57	6	0	80	0
A non-destroying checking process to find tiny cracks under 10 µm in ceramics is in practical use.	59	34	70	6	3	69	19	1	74	0
Supra-conducting materials are in industrial application at room temperature.	80	35	44	4	5	67	33	2	69	1
Machine tools will be developed which are so constructed that the workpieces are not de-formed by heat.	16	38	85	1	2	28	47	1	69	1
Manufacturers of durable consumer goods are obliged by law to take back and dispose of their products at the end of their operating life, where-by a recycling system with planning, production, collecting and recycling or re-use is in common use by means of which an almost complete material closed loop is achieved.	6	4	99	11	0	20	84	0	33	0
Factories to manufacture new materials are built in space, in which both high vacuum and weight-lessness are utilized to advantage.	96	3	10	4	11	90	14	9	23	0

7.5   Chemistry and Materials

The 31 topics of the field Chemistry and Materials that can be compared directly are rated similarly by the Japanese and German experts. Even the expertise for the single topics is similar: when rated high in Germany, it is also high in Japan, when rated low in Germany, it is also low in Japan. This shows a similar specification in the countries' knowledge, or that there are topics for which until now no knowledge is available in either country, whereas in others, the knowledge is really international. The only exception for the expertise is a topic from marine sciences *(processes designed to produce fuels from microorganisms and algae are in common use, with the result that the proportion of these fuels (e.g. alcohol) reaches 10 % of world production)*, which do not belong to the German research priorities. The other topic *(a new refining technique will be developed which makes the mining of titanium as cheap as that of aluminium)* is not that easily explicable. For this topic, the Japanese experts rate their R&D level significantly higher than the German experts think the Japanese are. Especially in the field of Chemistry, there are significant differences (table 7.5-1), in the first case of the table even in the assessment of the American R&D level.

Most of the topics are estimated to be very important for the economy. Nevertheless, for some of the topics, there are differences in the estimations (table 7.5-2). In the first and the second topic, there is also a difference in the estimation of the relevance for the enlargement of human knowledge, which is rated higher by the German experts than by the Japanese. For an explanation, further research would be necessary. But the importance for the knowledge enchancement does not reach the same high level as the one for the economy.

On the other hand, the estimations concerning the time of realisation are quite similar. But again, the need for a better education in Japan is pronounced for the whole sample of comparable topics. And again, more environmental follow-up problems are expected for every single topic.

Table 7.5-1: Assessment of the R&D level in topics with significant differences (grey boxes stress the difference, in per cent, for calculation see chapter 2.5)[13]

Chemistry & Materials	Germany R&D level					Japan R&D level				
	USA	Japan	Germany	other EU-country	other country	USA	EU	former USSR	Japan	others
Self-regenerating polymers will be developed.	92	42	30	7	3	65	15	0	50	0
Combined systems to treat organic matter in refuse by means of methane fermentation technology and waste incineration are in practical use.	12	18	96	10	4	31	38	2	70	0
Non-breaking control systems to examine tiny cracks of less than 10 µm in ceramics are in practical use.	39	48	77	8	0	69	19	1	74	0
A supra-conducting substance will be developed whose cracking temperature is at normal temperature.	92	49	34	8	7	76	40	6	71	1
A technique will be developed which enables the magnetic flow in supra-conductors to be controlled.	97	36	47	5	0	79	36	5	63	0
According to a technique applied in STM (Scanning Tunnel Microscope) methods will be developed to regenerate surface flaws on crystal silicon dice and seal out dirt.	91	46	42	4	6	85	34	0	77	1
Composite materials of wood and non-wood are used, which retain the special material properties of wood and are easy to form and work, as well as having great strength and universal applications.	47	22	73	14	10	49	19	0	67	1
Storage cells based on photochemical hole burning (PHB) with high storage density are in widespread use.	82	57	40	2	6	81	30	2	73	0
The percentage of vehicles fueled by hydrogen which are equipped with hydrogen-storing metal alloys is over 10%.	55	39	90	5	5	66	30	2	74	0
A new refining technology will be developed which makes the mining of titanium as cheap as that of aluminium.	84	26	21	6	15	67	16	7	57	1

---

13 Note that the R&D level cannot be compared directly.

Table 7.5-2: Topics with differences in the importance for the economy[14] (grey boxes stress the difference)

Chemistry & Materials	Germany		Japan	
	Knowledge enhancement	Economy	Economy	Knowledge enhancement
A technique will be developed, by which nano-scale particles are sintered at temperatures of 800°C in a solid state to produce high-performance materials (e.g. ceramic materials) on a SiC or Si3N4 basis.	20,5	97,5	85,6	8,9
Non-breaking control systems to examine tiny cracks of less than 10 μm in ceramics are in practical use.	15,2	97,3	77,8	8,6
Solid electrolytical fuel cells with performance of several 10 MW are in general use not only in regional combined heat and power generation (heat and electricity supply) but also in decentralized electricity power plants.	10,0	94,2	45,3	4,4
Rechargeable polymer batteries with a volume-specific energy density of 400 Wh/l are in practical use (presently used NiCd batteries: 180Wh/l).	12,3	93,4	78,2	5,3
Synthetic membranes which posses an active transport mechanism analogous to biological membranes are applied in practice.	49,6	91,5	31,7	7,3
Composite materials of wood and non-wood are used, which retain the special material properties of wood and are easy to form and work, as well as having great strength and universal applications.	7,0	91,0	52,9	0
Large-surface, amorphous silicon solar cells which attain a degree of efficiency of more than 20% in energy transformation, are in general use.	5,0	87,6	65,3	6,3
A process to dissociate water by solar radiation is in practical use.	21,0	81,5	46,9	14,7
Processes designed to produce fuels from microorganisms and algae are in common use, with the result that the proportion of these fuels (e.g. alcohol) reaches 10 % of world production.	5,3	80,5	57,9	1,4
The percentage of hydrogen-propelled vehicles which are equipped with hydrogen-storing metal alloys is over 10%.	3,5	72,3	52,3	2,6

[14] Please note that only two categories are given for each country. Therefore, the percentages do not add up to 100 %. The table only shows the difference of high versus low importance.

7.6   Health and Life Processes

In this field, 36 topics can be compared systematically. The German experts rated their expertise relatively low, because they are – especially in medicine – specialised to such an extent that they do not regard themselves as experts concerning the borders of their focused disciplines. But the Japanese experts were not better and assessed their expertise similarly.

The field of Health and Life Processes is not a surprising one. What is surprising is the difference in the estimation concerning the importance for the enlargement of human knowledge (table 7.6-1), which is assessed to be much higher in Germany than in Japan. As it seems quite logical that the topics might improve knowledge, the Japanese experts do not see major improvements. They also rate the overall importance as rather *medium*.

To explain this difference, the following hypotheses were attempted:

- The sample of the expert groups concerning this field is very different, e.g. in Germany more doctors working in scientific research versus practitioners in Japan. But this cannot be proven as the sample in Japan includes a majority of experts from universities (64%) who are research-related (66%). This could only be checked by interviewing the single experts - which is of course impossible.

- The Japanese understanding of medicine and health is more target-oriented towards "curing" than "research", and the other categories fit more into this understanding. This cannot be proven, but as multi-choice without limitation was possible in the questionnaire, this does not explain the low number of crosses from the methodological view but more from the "content" view.

- Topics concerning the nervous system (7.6-1) get relatively high percentages. One reason might be that the experts from the large research programmes in neuro-sciences evoke a bias in the sample. Another reason might be that the propaganda concerning these large programmes (neuro-sciences were politically declared as key sciences) influenced the experts, so that they just re-stated the political view. But this could also only be clarified in detailed interrviews.

Looking at table 7.6-2, it is obvious that those topics which concern "hard products" which can be sold (without services) get low rates for the importance concerning the enlargement of human knowledge. The degrees for topics concerning "processes", "mechanisms", "methods", "therapies" are higher. Certainly, products are not pure basic research but the German experts seem to acknowledge much more that a huge knowledge-input is necessary to develop these "products" for a market. Maybe the Japanese experts regard this more technically as "engineering tasks".

Table 7.6-1: Differences in the importance for the enhancement of human knowledge and importance for the economy[15] (in per cent, see chapter 2.5, grey boxes stress the difference)

Health & Life Processes	Germany Importance for		Japan Importance for	
	the enhancement of human knowledge	the economy	the economy	the enhancement of human knowledge
A technique will be developed by which a number of neurons can simultaneously be experimentally observed and analysed for hours.	96,8	7,4	21,2	76,3
The mechanisms for the formation of neuronal networks is explained on the molecular level.	96,3	16,8	9,2	82,3
The reciprocal regulating mechanisms of the immune system, the nervous system and the endocrine system are understood, so that the findings can be put to therapeutic use.	83,7	25,2	12,7	35,2
The pathogenesis of Alzheimer's disease is understood.	75,2	25,5	31	23
Electrical leads which can be connected to nerve and brain cells, e.g. for artificial eyes, are developed.	70,6	26,5	36,4	31,8
Artificial intelligence which connects ICs with living cells in hybrid form are developed.	69,9	43,0	54,8	42,9
A method is developed by which severed nerves of the central nervous system can be connected together again.	68,1	14,2	27,3	32,7
A therapy will be developed to completely cure schizophrenia.	67,9	17,0	29,6	29,6
Implantable and fully functioning artificial ears are clinically inserted, with which deafness due not only to defects of the sound conducting organ, but also of the sound pick-up organ can be conquered.	62,0	25,0	48,6	16,2
An effective therapy is developed for Alzheimer's disease.	56,3	42,1	33,3	23,6

15 Please note that only two categories are given for each country. Therefore, the percentages do not add up to 100 %. The table only shows the difference of high versus low importance.

Table 7.6-2: Other topics of the health sector (with differences in the impor-
tances in per cent, see chapter 2.5, grey boxes stress the difference)

Health & Life Processes	Germany		Japan	
	Importance for			
	the enhancement of human knowledge	the economy	the economy	the enhancement of human knowledge
Processes for the prognosis of the functions from the spatial structure of glyco proteins are developed.	91,1	50,0	55	78
The processes of physical ageing are understood.	87,3	26,1	33,3	35,7
Artificial membranes are developed whose functions are similar to biological membranes (active transport, energy conversion, information transmission).	81,9	67,2	68,7	22,7
The mechanisms of tolerance reaction are understood, so that transplantations of the inner organs is mastered.	81,5	38,9	16,9	16,9
The signal transmission in the developing cancerous cells will be so regulated that therapies are in widespread use by which an induction to redifferentiate or normalize cancer cells takes place.	73,9	25,2	16,1	22,4
Processes to produce and provide various blood corpuscles (for artificial blood) are in practical use.	71,8	67,0	38,8	20,4
Techniques to regenerate organs via the reproduction of own cells are established and applied clinically.	71,3	57,4	22	10,7
Technical systems will be developed which can reproduce themselves, like living organisms.	70,5	55,8	64,7	37
An early diagnosis of the defensive reactions to organ and tissue transplantations is in practical use.	69,4	21,5	22,9	18,6
Therapeutic methods to remove viruses from the blood are in general use.	68,0	46,6	38,1	22,2
Processes to differentiate linear fibre germ cells for the cardiac muscle are used clinically for the treatment of coronaries.	67,9	28,3	28	24
A technique for the long-term conservation and cultivation of organs will be in general use.	66,0	54,4	28,1	7,2

Biosensors will be applied which utilise e.g. antibodies.	62,2	75,6	42,9	3,2
Methods based on genetic analysis to predict the individual risk of illness for genetically co-determined diseases (e.g. cancer, high blood pressure) are very common.	61,8	26,3	26,3	25,3
Medicines are in widespread use which can recognize their targets (e.g. tumor cells) and zoom in on them (missile drug).	60,6	47,9	26,7	1,7
Effective radiation sensitizers are developed for cancer therapy.	60,2	32,0	41,2	14,7
Artificial organs (pancreas, kidney, liver, lung etc.), in which human cells and tissue are integrated, are in practical application.	55,6	45,2	35,4	4,1
Medicines for the treatment of acute viral hepatitis are in widespread use.	53,8	56,4	37,8	14,9
A method, by which the stage and extent of the disease focus of arteriosclerosis can be diagnosed, is in clinical use.	48,0	36,6	27,3	16,7
Preventive measures to hinder congenital deformities which occur in the embryonal and perinatal development phases are in use.	47,6	6,3	22,4	17,2
Processes by which blood samples for biochemical analysis are not only collected but can be taken on the body's surface without direct penetration, are in clinical application.	46,2	68,2	46,5	15,5
Ultramicro-biosensors on the basis of biochemical reactions are in practical medical application.	45,5	72,7	42,1	1,4
An effective insuline preparation will be developed which can be administered orally.	36,1	55,5	44,4	12,5
Remote operation systems using virtual reality technology are in widespread use.	25,9	73,2	30,1	3,2

In the case of the importance for the economy, the picture is more differentiated as there are topics with a higher importance for the economy in Germany (table 7.6-1): *ultramicro-biosensors based on biochemical reactions are used for medical purposes; medicines are widespread which recognise their targets (e.g. tumor cells) and can zoom in on them (missile drug); medicines against acute viral hepatitis are widespread; remote operations systems using virtual reality techniques are in common use; the mechanisms of tolerance reaction are understood, so that transplantations of the inner organs have been mastered; techniques to regenerate organs by reproducing own cells are established and in clinical application; biosensors are applied which use e.g. antibodies; a technique for the long-time conservation and cultivation of organs is in general use; processes to produce and provide diverse blood corpuscles (for artificial blood) are in practical use).* An explanation cannot be given.

The estimation of the time of realisation does not differ very much in this innovation field. The major difference of eight years in the median can be observed for *techniques for the regeneration of organs by the reproduction of own cells will be established and in clinical application.* Here, the German experts are more optimistic (2011-2014-2018) than the Japanese (2016-2022 after 2025). Both groups expect a more long-term perspective.

There are no surprises concerning the R&D level. Both countries think the USA are the leading country in the world in the field of health research. What is striking to a certain extent is that the German experts rate the level of R&D in Japan as high as the Japanese experts do for sensors *(ultramicro-biosensors on the basis of biochemical reactions in medical application; processes by which blood samples for biochemical analysis are not only collected but can be taken on the body's surface without direct penetration).* In general, and because of the statistical incompatibilities, the Japanese assess their R&D level higher than the German experts do. The same phenomenon applies for artificial intelligence and other topics concerned with communication technologies or new electronics *(artificial intelligence which connects ICs with living cells in hybrid form; electrical leads which can be connected to nerve and brain cells, e.g. for artificial eyes; use of remote surgery systems with virtual reality* and *robots to assist in the care of the sick, the elderly and persons with serious physical disabilities or mental disorders).* Do the German experts overestimate the Japanese research? Or are the Japanese experts just humble or do not know exactly the R&D level in the own country? This is something that has to be checked in detail further on.

It is also no surprise that the Japanese experts call for better education in all themes comparable in this field, too. In Japan, huge ethical problems are expected, the expectations for social, cultural and ethical problems in Germany are even higher. For some topics, also severe security problems are expected in Japan, but not as much in Germany (e.g. for the different *biosensors*).

7.7   Agriculture and Food

In the innovation field Agriculture and Food, 20 topics can be compared. The expertise for this sample is slightly higher in Japan than in Germany but does not seem to have had an influence on the estimations which differ from topic to topic in this field. Especially interesting are the differences concerning the importance for the economy and for the enhancement of human knowledge (table 7.7-1). The importance for the economy is rated to be much higher in Germany than in Japan in most cases.

Table 7.7-1: Differences in the importance for the economy and for the enhancement of human knowledge[16] (in per cent, see chapter 2.5, grey boxes stress the difference)

Agriculture & Food	Germany Importance for the economy	Japan Importance for the economy
The cloning of prize-winning, high-performance cattle by core transplantation is practised.	95.0	46.1
When a data collection and storage system (DCS) using satellites and markers is implemented, which collects data, analyses and communication over wide areas continuously, completely and simultaneously, the service network for information about fisheries and the marine situation is completed.	93.9	36.5
Cell fusion and gene technology will make possible the cultivation of new breeds of fish which are very suitable for fish farming due to their strong resistence to disease and fluctuations in water temperature.	93.8	56.3
In order to achieve certain breeding goals (resistence to disease, fertility) in domestic animals, gene transfer to fertilised eggs or to early mammal embryos is practised.	91.3	44.4
Simulation systems will be developed for agricultural management with which the latest agricultural technologies can be applied.	85.9	60.6
Foodstuffs are widespread in which unicellular organisms like microorganisms or algae serve as raw material.	82.4	40.5
Techniques are widespread, e.g. using microorganisms, which enable earth-bound phosphorus to be absorbed by cereals.	79.8	22.4
Plants which are specially cultivated for resistance to drought and salt and provide barriers to desertification are in practical use.	78.3	25
Biological control systems are widespread which offer protection against disease and vermin through biological pesticides (natural microbial enemies, pheromones etc.).	74.3	26.2
Biodegradable packaging manufactured from renewable raw materials are in common use.	70.5	50.8

---

16 Please note that only two categories are given for each country. Therefore, the percentages do not add up to 100 %. The table only shows the difference of high versus low importance.

A system to utilise marine organisms and their environment is achieved which can keep the balance between the exploitation by the fishing industry and the habits of fish shoals under the prevailing biological and ecological conditions.	61.5	25.7
After the mechanisms of forms and functions of the ecosystems are understood, rational monitoring and exploitation procedures for rainforests, including the presently existing life forms, will be implemented in tropical regions.	50.7	10.6
The use of transgenetic animals, into which genes that hamper or prevent the defensive reactions in xenotransplantations were transplanted, is widespread for the transplantation therapies of inner organs.	50.0	37.2

In three cases (cell fusion and gene technology for new fish, cloning, and xeno transplantation), there are significant differences observable in the estimation of the importance for the human knowledge. In all three cases, this is rated higher by the German experts, in the first two cases on a low level, concerning *xeno transplantation* on a level of medium importance.

The time of realisation is estimated quite similarly in both countries. The major differences concern genetically modified plants: *genetically modified plants which are able to bind large quantities of CO_2 will be developed and utilised in environmental protection* (time: 2012-2018-2024 in Germany, 2009-2013-2018 in Japan) and *genetically modified plants and microorganisms resistent to NO_X or which reduce the NO_X content in the air, are applied in environmental protection* (time: 2016-2021-after 2025 in Germany with 22.9 % saying never, 2008-2012-2017 in Japan, only 0.7 % say never).

Like in other fields, in general the USA are regarded as the leading country in the world. But especially in the marine sciences and fishing, the German experts think that the Japanese are the best (for: *cell fusion and gene technology will make possible the cultivation of new breeds of fish which are very suitable for fish farming due to their strong resistance to disease and fluctuations in water temperature; a system to utilise marine organisms and their environment is achieved which can keep the balance between the exploitation by the fishing industry and the habits of fish shoals under the prevailing biological and ecological conditions; and foodstuffs are widespread in which unicellular organisms like microorganisms or algae serve as raw material*). Interestingly, the Japanese do not think that they are that strong (they give themselves only half of the percentage), although for nearly all other topics, the German experts do not think that they are as advanced in R&D as the Japanese rate themselves. This seems to be a clear prejudice: as fish is a more important factor for food production and the economy in general, the experts extrapolate a better R&D position for Japan, too.

Unsurprisingly, the Japanese call much more often for better education than the German experts do. In two cases, the German experts demand a different regulation much more often than the Japanese persons (*a method whereby the environmentally protective functions of forests can be assessed quantitatively, so that monitoring techniques can be widely used which preserve these functions and simultaneously allow the resources of the woods to be exploited* and *biodegradable packaging manufactured from renewable raw materials is in common use*). The different regulations that have an effect on the topics have to be clarified in detail.

As in most fields, in Germany, there are more follow-up problems expected than in Japan. This concerns mainly environmental problems but also social or cultural problems. An exception are the *foodstuffs are widespread in which unicellular organisms like microorganisms or algae serve as raw material*, for which the Japanese experts expect slightly more (and not too many) environmental problems.

7.8   Environment and Nature

In this field, 14 topics can be compared directly. But there are also some similar topics which might be comparable in a more qualitative way. The expertise for the 14 selected topics is similar in Japan and Germany.

There are some topics, for which the results are striking, especially in the estimation of the importance (table 7.8-1). Whereas the German experts regard *technologies for the greening of desert areas* as of medium importance for the economy, the Japanese do not agree in this point. The same is true for the *re-use of water*. Maybe the Japanese experts expect high costs for this system so that the net effect is regarded as small and therefore the topic is not that interesting for the economy.

The third major difference can be observed for *predictions of changes in global ocean currents for flood-endangered coasts*. The German experts assess this topic as relevant for the economy to a larger extent than their Japanese colleagues. As Japan has more access to the sea and is therefore more often and severely affected by floods than Germany, this cannot easily be explained. Is the danger for the economy ignored? Or will the impact really be small?

Table 7.8-1:   Differences in the importance for the economy and the enhancement of human knowledge (in per cent, see chapter 2.5, grey boxes stress the difference)

Environment & Nature	Germany		Japan	
	Importance for			
	Knowledge enhancement	Economy	Economy	Knowledge enhancement
The combination of biological systems and new technology facilitates the wide spread of marine farms which are so well integrated that they are profitable and do not negatively influence the marine environment.	14	83	60	3
A method to predict changes in global ocean currents before flood-endangered coasts is in practical use.	21	79	27	22
Advances in the decentral treatment of waste water which will improve the possibilities of water re-cycling not only for industry but also for muncipal use are widely applied.	3	75	28	1
Technologies to green desert areas are in worldwide use to prevent further desertification.	14	67	28	2
Combined systems to treat organic refuse using methane fermentation techniques and incineration are in practical use.	4	58	32	4
The interactions between the phenomena occurring prior to earthquakes and the behaviour of animals are understood and used in earthquake prediction.	64	53	5	24
A remote sensing technology which provides exact information, e.g. about water temperatures, ocean currents and the chlorophyll concentrations in up to 200 m water depth using satellites, is in practical use.	36	48	12	25

The same phenomenon concerns the topic *earthquake prediction on the basis of animal behaviour*. This method is applied in China to a huge extent where the system of seismometres and sensors is not as widespread as in Japan just because the country is larger. The Japanese experts regard their measuring system as the better one, although until now it was not more successful than the Chinese approach (Cuhls 1998). Therefore, the Japanese experts might have assessed the importance of the *system to observe the behaviour of animals* as irrelevant for the economy as well as irrelevant for the enlargement of human knowledge. On the other hand, they might have judged the *possibility to predict an earthquake* and use the prediction for measures – and this has an impact on the economy. But as the Kôbe earthquake showed, the impact of the earthquake as such can be disastrous for the whole economy.

Concerning the estimation on the time of realisation, it is most striking that the Japanese experts are much more pessimistic than their German counterparts concerning the *worldwide CO_2 emissions*. The German experts think the emissions can be reduced by 20 % compared to 1990 by the year 2015 (range from 2010 to 2020) and only 10.9 % of the experts say "never", whereas the Japanese do not expect this reduction until 2022 (range from 2016 to 2027). 31.7 % say "never", this is nearly one third of the experts.

The Japanese experts regard their own country as leading the world, in most cases of the sample more often and therefore better than the USA. There are three exceptions, for which the German experts to a very high percentage regard the Japanese R&D as the best in the world but the Japanese are much more modest:

1. *The combination of biological systems and new technology facilitates the wide spread of marine farms which are so well integrated that they are profitable and do not negatively influence the marine environment.* (94 % of the German experts rate Japan as leading, but only 76 % of the Japanese).
2. *The cultivation of microorganisms in the sea and on the seabed for technical purposes is widespread* (94 % of the German experts rate Japan as leading, but only 65 % of the Japanese).
3. *The interactions between the phenomena occurring prior to earthquakes and the behaviour of animals are understood and used in earthquake prediction.* (89 % of the German experts rate Japan as leading, only 48 % of the Japanese). This topic was discussed above as more differences can be observed here.

As in most of the innovation fields, the Japanese experts call for better education more often than the German experts. Especially in Environment and Nature, education seems to be no problem in Germany. On the other hand, a change in regulation is called for in Germany for the following topics whereas in Japan, a significantly lower percentage of the experts seek a change in regulation:

1. *Combined systems to treat organic refuse using methane fermentation techniques and incineration are in practical use.* (G: 73 %; J: 27 %)
2. *A refuse derived fuel system (RDF) is in widespread use to produce electricity through refuse incineration.* (G: 79 %; J: 58 %)
3. *Short-lived containers and packaging will be manufactured mostly from biodegradable and completely mineralisble synthetics.* (G: 73 %; J: 29 %)
4. *Biotechnological or chemical-physical processes are developed with which oil pollution caused by tanker accidents can be removed effectively and in an environmentally friendly manner.* (G: 49 %; J: 10 %)

Although in general the topics were regarded as important for the solution of environmental problems (only asked for in Germany), in Germany more environmental follow-up problems are expected for most of the topics. This fear is not as high in Japan where around 50 % of the experts expect environmental problems. In some cases, also social-cultural or ethical problems are expected. In these cases, the German and Japanese experts agree on the assessment.

7.9   Energy and Resources

Energy and Resources is one of the innovations fields where many topics are international in nature and can be compared. 32 topics are equivalent in the Japanese and the German questionnaire. Also the expertise is very similar for both samples.

For many of the topics selected, the importance for the economy is rated much higher in German than in Japan (table 7.9-1). Especially striking is the *fast breeder system*, which gets twice the rating in Germany, although Japan is definitely the leading country in the world from the experiential point of view, and although in Germany, nuclear energy is criticised very much and will be abandoned (see also Cuhls 1998).

The differences in the time of realisation are rather minor ones. Only for a few topics, a difference of five or more years can be observed in the expectations (table 7.9-2).

The Japanese experts regard themselves more often as the leading country in the world than the German experts think the Japanese are. This is mainly the case in topics that have to do with *coal*, and for the *sea wave power generation*.

Generally, better education is called for in Japan, which is not regarded as a problem in Germany, although the new discussion is coming up that in the next few years there will be a lack of experts for nuclear power plants (from energy generation to recycling) for the future because many people will soon be retiring, and it is no longer attractive to study this field because the government decided to withdraw from nuclear power generation. This is not (yet?) reflected in the Delphi study.

For many topics in the field of Energy and Resources, environmental follow-up problems are expected. Although the percentage of those persons who expect these problems is generally much higher in Germany than in Japan, follow-up problems are mentioned, there, too. But there are topics which on the one hand will be important for the economy or the solution of environmental problems but will cause new problems on the other – so the German experts.

Table 7.9-1:   Differences in the importance for the economy (ranked from the German perspective, in per cent, see chapter 2.5)

Topic	Germany	Japan
Prospecting technology is developed with which the mineability prospects for ore beds can be estimated with hardly any trial drilling.	98	27
Solution mining is in practical application with which metal deposits deep in the earth such as e.g. copper pyrite or sulphide minerals with lead or zinc, can be dissolved and pumped up from the surface.	97	32
Non-electrolytical reducing processes to obtain metallic aluminium are applied in practice.	95	47
A direct-current (DC) transmission over great distances with a voltage of 1,000 kV is in practical application (today ca. 6000 km can be reached with 600 kV).	93	33
Combined gas and steam turbine electric power stations with extremely high degrees of efficiency (inlet temperature over 1500°C; today ca. 1400°C) are in common use.	92	42
Electricity transmission or distribution networks utilizing supra-conducting cables are in practical use.	89	49
Plants using gasified coal to produce electricity are in common use.	88	40
In-situ gasification of of coal seams will be carried out.	85	28
Technology to hydrate coal is utilised commercially.	81	32
The use of biotechnological processes to dissolve and extract metals from ore and waste is widespread.	81	31
Solid matter electrolytical fuel cells with a capacity of several 10 megawatts are in common use, not only in regional combined power and heat generation (heat and electricity supply) but also in decentral electricity power stations.	78	45
The production of methane and methanol fuel from coal and biomass, utilising non-fossile (i.e. obtained electrolytically from water) hydrogen is in practical use.	75	28

Power stations on the basis of coal gas operated high-temperature fuel cells with a capacity between 200 to 300 MW will be put into operation (capacity of present-day pilot plants: 2 - 100 KW).	74	32
Supra-conducting energy-storage systems with capacities similar to those of pumped storage power stations (one million kWh) are in practical application.	74	38
A fast breeder that is integrated in a nuclear fuel cycle is in common use.	72	36
Gas turbines which can be run on hydrogen are in practical use for public electricity production.	72	28
Fuel cells based on solid matter polymers with combined power and heat generation are in widespread use in domestic housing.	72	42
A thermo-chemical process to produce hydrogen as an energy carrier is in practical use.	67	33
Highly efficient energy production from biomass (e.g. gas from plants, wood and straw) is in widespread use.	66	33
The biological production of hydrogen utilizing solar energy by means of intact organisms or biological subsystems is in industrial application.	62	33
Geothermal energy production utilising "hot dry rock" (HDR) is in practical use.	57	19

Table 7.9-2:   Differences in the time of realisation (the median as the base for calculation is marked in grey)

Energy & Resources	Germany time of realisation				Japan time of realisation			
	Q1	Median	Q2	never (in %)	Q1	Median	Q2	never (in %)
Wave power stations to utilise the energy of wave movements are in practical application.	2013	2019	after 2025	24	2004	2009	2015	3,4
The temperature differences in sea water are practically exploited for electricity production.	2019	after 2025	after 2025	34	2015	2020	2025	24,3
Power stations on the basis of coal gas operated high-temperature fuel cells with a capacity between 200 to 300 MW will be put into operation (capacity of present-day pilot plants: 2 - 100 KW).	2017	2023	after 2025	6,7	2014	2018	2022	6,7
Electricity transmission or distribution networks utilising supra-conducting cables are in practical use.	2015	2021	after 2025	5,9	2019	2025	after 2025	14,6
Solution mining is in practical application with which metal deposits deep in the earth such as e.g. copper pyrite or sulphide minerals with lead or zinc, can be dissolved and pumped up from the surface.	2011	2016	2020	5	2016	2021	after 2025	11,1
A direct-current (DC) transmission over great distances with a voltage of 1,000 kV is in practical application (to-day ca. 6000 km can be reached with 600 kV).	2008	2010	2014	0	2011	2016	2020	0,8

7.10   Construction and Living

In this field, 30 individual topics can be compared. There was a certain disappointment in Germany that the expertise in this field was relatively low. But it is similarly low in Japan, especially for specific topics. There is always the danger of not being able to identify the right persons to assess future topics or even to convince them to participate in a Delphi survey. But this result gives a hint that there are topics for which only limited expertise is available, even on an international level. It would be interesting to check with other countries like the USA.

There are some topics, for which the German experts rate a significantly higher importance for the economy, but generally on a medium scale (table 7.10-1). There are only two exceptions, where the estimations are much higher in Japan than in Germany: for the *marine cities* and for *systems to build cities in desert or polar regions*.

There are no major differences in the estimation of the time of realisation. The largest gap are five years for *a building technology is developed for buildings and structural and civil engineering, which takes maintenance and demolition into account at the planning stage.* The Japanese are also famous for their belief in technology. This is a prejudice. Even for those topics that were already exceptions in the importance for the economy, there are more persons in Japan estimating that the topics are not realisable: *progress in building in water makes it possible to construct "swimming cities"* (in Japan: 25.2 %, in Germany 14.3 % say "never") and *new technical systems will be used which enable cities to be built in desert or in polar regions* (in Japan: 32.2 %, in Germany 18.2 % say "never"). In this field, there is another, often criticised topic which many experts regard as never realisable: *building technology for skyscrapers (ca. 1,000 m high) will be utilised in Germany* (Germany: 25 %, Japan: 12.4 %).

For the R&D level, the results are more diffuse than in other fields. Exceptions here are again the *marine cities*, the *1,000 m skyscrapers*, and the *cities in desert or polar regions*. For the *marine cities* and the *skyscrapers*, the German experts think the Japanese are the leading or one of the leading countries in the world. Here, the Japanese experts give a more moderate self-estimation. For the *cities in desert or polar regions*, the Japanese experts rate themselves much better and claim to be the leading country in the world, much ahead of the Americans, too, whereas here, the German experts think the USA are the most advanced. Another over-estimation seems to be made by the German experts in case of the robots *(houses equipped with robots and appliances are generally available in which e.g. old or handicapped people can look after themselves unaided (cooking, bathing, toilet, entertainment etc.)*, for which the Germans assess Japan as the most advanced country, then the USA, whereas the Japanese experts think the European Union is more advanced than they are. This just shows mutual prejudices that can lead to the motivation to invest in certain fields of research. So one has to be very careful just to benchmark countries. More information is needed in these cases.

Table 7.10-1: Differences in the importance for the economy (in per cent, see chapter 2.5, grey boxes stress the higher percentage)

Topic	Germany	Japan
Progress in building in water makes it possible to construct "swimming" marine cities.	22	85
New technical systems are used which enable cities to be built in desert or in polar regions.	18	84
A building technology is developed for buildings and structural and civil engineering, which takes maintenance and demolition into account at the planning stage.	68	81
Remote monitoring and control systems are in general use to guarantee increased safety for supply lines (e.g. for water, electricity and gas).	83	53
Decentral energy supply systems for domestic housing utilising fuel cells or integrated energy production (heat and electricity) is in practical use.	61	30
Solar electrictiy-producing systems are very widespread in Germany which ensure economical and stable lighting for streets and tunnels.	62	29
An automatic sorting technology is in general use which separates normal domestic refuse according to degree of hardness, specific weight, dampness and colour into inflammable material, metal and glass.	61	35
A heating and cooling system with a heat pump based on the use of solar energy is in practical use in Germany.	71	40
Technology for measuring, steering and regulating will be used in practice with which intertior temperature in houses can be effectively controlled, utilising stored coldness and waste heat.	72	24
Materials for interior decoration in which sensors for room temperature and humidity as well as control elements for room atmosphere are integrated are developed.	61	25
The utilisation of end energy and refuse re-cycling is widespread in German communities.	59	32

Water works of medium capacity which recycle waste water into industrial water, are decentrally distributed in the individual areas of the large cities.	51	21
A monitoring system for advance warnings, predictions and evacuations on the basis of local weather predictions is widespread in Germany, so that personal injuries in catastrophes, such as e.g. in rivers or on the streets, can be drastically reduced.	48	20
A technology to fight fires and rescue people in fires in high-rise buildings will be developed.	44	28
Modular houses in which the allocation of rooms and furniture/fittings can easily be adapted to the life phases of the inhabitants or e.g. a change in generation, are widespread.	39	25
The use of robots to fight fires, which can recognise and rescue people, is very widespread in Germany.	34	20

The measures that should be taken show a diffuse picture (table 7.10-2). In Germany, this is the field in which many experts call for better education, the Japanese experts call for this even more often. For some single topics, regulation problems seem to exist, more in Japan than in Germany. Interestingly, in Japan, the whole R&D system is often regarded as the major problem.

In most cases of this field, the German experts expect more environmental follow-up problems than the Japanese experts, who do not ignore them but mention them more selectively. Especially for the *cities in polar or desert regions*, the Japanese expect more security problems. On the other hand, the German experts estimate that *new building materials which consist of high-polymer fibres or ceramic, will be developed and in widespread use in construction , as well as for bridges and dams* (Germany: 71 %; Japan: 32 %) and *through the development of a steel glue with high performance adhesive qualities steel construction will be considerably rationalised* (Germany: 79 %; Japan: 40 %) will evoke more security problems than their Japanese colleagues expect.

In the case of *robots (houses equipped with robots and appliances are generally available in which e.g. old or handicapped people can look after themselves unaided (cooking, bathing, toilet, entertainment etc.)* and of *modular houses (houses in which the allocation of rooms and furniture/fittings can easily be adapted to the life phases of the inhabitants or e.g. a change in generation, are widespread)* especially many social and cultural problems are expected in Germany (94 % and 90 %), but not many ethical problems in Japan (42 % and 23 % respectively). This means that the experts really differentiated between the socio-cultural dimension (which included ethics according to the definition) in Germany and the pure ethical dimension asked for in Japan.

Table 7.10-2: Differences in the measures "better education" and "change in regulation" to be taken (in per cent, see chapter 2.5, grey boxes stress the higher percentage)

Construction & Living	Germany		Japan	
	Measures			
	Education	Regulation	Education	Regulation
A building technology is developed for buildings and structural and civil engineering, which takes maintenance and demolition into account at the planning stage.	56,9	55,6	39,1	34,8
Important parameters for construction plans and building planning, like soil properties, geological structure and climatic conditions are collected in Germany in widespread, uniform databases.	43,7	42,3	29,6	12,4
Compact used water recycling systems are in widespread use in Germany, which on a biotechnological basis very effectively dispose of substances with low solubility or pollutants.	12,3	56,1	44	24,8
Modular houses (houses in which the allocation of rooms and furniture/fittings can easily be adapted to the life phases of the inhabitants or e.g. a change in generation) are widespread.	40,9	40,9	17	54,8
Space stations are ready for operation, in which untrained personnel can spend a longer time (longer than one year).	15,6	0,0	65,3	0
Building materials and systems are developed which are capable of self-diagnosis and self-repair.	29,8	8,8	56,9	7,7
Through the development of a steel glue with high performance adhesive qualities steel construction is considerably rationalised.	26,5	8,2	52,4	18,3
Building technologies for skyscrapers (ca. 1,000 m high) are used in Germany.	9,1	9,1	51,9	7,8

A monitoring system for advance warnings, predictions and evacuations on the basis of local weather predictions is widespread in Germany, so that personal injuries in catastrophes, such as e.g. in rivers or on the streets, can be drastically reduced.	17,0	10,6	47,7	11,6
New technical systems are used which enable cities to be built in desert or in polar regions.	12,8	6,4	25,1	56,1
New, universal tunnel systems are widespread in inner cities in which the cables for radio, pipes for vacuum cleaning plants, for transporting goods and for heating and air conditioning are laid.	9,8	29,4	12,3	68,2
Progress in building in water makes it possible to construct "swimming" marine cities.	16,3	4,7	20,3	48,3

7.11 Mobility and Transport

The expertise for the 17 comparable topics was regarded as disappointing in Germany. But looking at the topics in detail, the expertise is even worse in the Japanese sample and differs in the same manner as the German expertise from one topic to the other.

The importance of the topics for the economy is generally regarded as higher in Germany than in Japan. There is only one exception, the *super conducting magnet train with a speed of 500 km/h*, which the Japanese experts regard as much more important for the economy than the German experts (Japan: 94 %; Germany: 76 %). Most of the topics are not important for the enlargement of human knowledge – estimated in Japan as well as in Germany.

The time of realisation does not differ very much, too. Exception here is firstly, again the *super conducting magnet train with a speed of 500 km/h*, which is regarded as never realisable by 14 % of the German and only 3.5 % of the Japanese experts. Those experts who regard it as possible estimate a similar time frame (Germany: 2009-2013-2017; Japan: 2008-2012-2016).

The second exception are *flying boats (hydrofoils, airfoils) will be in everyday use in scheduled services between towns or outlying islands, due to the advances made in new, salt-resistant materials and motors)* which are estimated to be used earlier by the German experts (2004-2008-2011) than by the Japanese counterparts (2012-2015-2018). 17.3 % of the Japanese experts even rate them as impossible (0 % in

Germany). The reason might be that these kind of boats already exist but not in *practical use* between towns or small islands. The optimistic view of the German experts is nevertheless striking as there are no small island or towns in this country that could be connected at all.

The last mentioned topic is also one of those for which the German experts rate the Japanese as the most advanced in the world, an estimation that is not shared by the Japanese themselves. And both countries rate neither Germany nor the whole EU as competent or advanced in this field. Maybe the German experts just do not know enough about the real situation of an island.

Táble 7.11-1:   Differences in the regulation measures to be taken (automobiles), (in per cent, for calculation see chapter 2.5)

Topic	Germany	Japan
Silencing equipment is used, which is installed on the roads so that traffic noise is absorbed high-energetically to reduce the noise level.	82	9,2
90 % of car parts and materials can be recycled or re-used.	81	45
Extremely economical cars with 30 % less petrol consumption than today's cars, due to basic technology developments like the introduction of new materials which make cars stronger and lighter and more combustion-efficient motors, are widespread.	66	19
Emission technologies are utilised e.g. diesel catalytic converters, particulate traps, lean-NO catalysts or very precise combustion technology, so that the noxious components of the emissions are reduced to 1/10.	63	26
By means of improved engines, transmission, silencers, tyres and road surfacings the noise of trucks is reduced to the level of today's cars.	58	26
Electric cars are in common use in which highly efficient fuel cells are utilised for energy conversion.	37	16
Cars and machines which use hydrogen as fuel instead of petrol or alcohol, are in general use.	31	23

For topics that have to do with cars, German as well as Japanese experts claim for themselves to belong to the leading country in R&D in the world, e.g. for *extremely economical cars with 30 % less petrol consumption than today's cars, due to basic technology developments like the introduction of new materials which make cars stronger and lighter and more combustion-efficient motors, are widespread; cars and machines which use hydrogen as fuel instead of petrol or alcohol, are in general use; emission technologies are utilised e.g. diesel catalytic converters, particulate traps, lean-NOx catalysts or very precise combustion technology, so that the noxious components of the emissions are reduced to 1/10* or *driver-suupport systems are used on the streets, which receive the information required by the driver and give the driver warning or intervene in the driving process.* But also in the case of the *magnetic train*, both experts groups claim their country to be the leading one in the world.

Concerning the measures to be taken, the Japanese experts call for better education, again, whereas this seems to be no problem in Germany at all. What is especially interesting is the attitude towards cars and emissions that can be interpreted from the measures to be taken. Whereas the German experts call for a change in regulation, this is not as often mentioned by the Japanese experts (table 7.11-1). But in exactly these cases, the Japanese ask for more funding of R&D. This is not directly comparable to the German categories that have to do with financing (financing by third parties and R&D infrastructure), but as the German experts do not mention the financial measures in these cases it can be concluded that whereas the German experts trust more in regulation to solve the problem, the Japanese experts seem to regard money for research as the better solution.

Congruent to these findings, the German experts expect generally more follow-up problems for the environment but also more security problems, especially in the case of *alcohol or hydrogen cars* as well as for *electric cars with fuel cells*.

7.12   Space

In the innovation field that deals with inner and outer space, the expertise for the comparable Delphi topics is very similar. But some differences concerning the importance of the topics can be observed (table 7.12-1). Generally, the German experts regard the topics as more important for the enhancement of human knowledge. Especially topics on satellites or radar get relatively low assessments in Japan, and more medium ratings in Germany. These are also the topics that are regarded as much more important for the economy in Germany. On the other hand, the *multidirectional radio technology* and the *research facilities on the moon* are regarded as slightly more important for the economy in Japan.

Although the time horizon of many of the topics is rather long-term, there is an international consensus on the time of realisation. Both the German and the Japanese expert group estimate very similar realisation times and a similar R&D level. Both groups think the USA is the leading country in the world. There is only one exception: for topics that are concerned with satellites, the German experts do not regard the Japanese as the leading in the world (they give rather medium estimations) but the Japanese give themselves better notes for *by using a data collection and storage (DCS) system employing satellites and measuring buoys that ensures long-term, complete and simultaneous data collection, performance analyses and communication over large areas, the service network for information on the fishing and the state of the seas are completed; movements in the earth's crust are measured to within less than 1 cm of error by very long baseline interferometers (VLBI), satellite-assisted laser radar, inversion laser distance measurement or LIDAR with synthetic apertures so that the the prediction of earthquakes can be improved* and *a system is in practical use that can exchange signals via satellites from many terrestrial stations in multiplex operation.* Especially for the *multi directional radio technology* already mentioned, the German experts give very low assessments whereas the Japanese regard themselves much better, even better than the USA.

One reason can always be that the German experts just do not have the right expertise, but especially for these topics they claim a higher expertise than for the other topics in the field. A second reason might be an over-estimation on the Japanese side – as they have sophisticated plans for space and satellite activities, they claim to be the leading country in the world in order to get sufficient funding for the future, too. This fits the result that a high percentage of the Japanese experts call for more R&D funding, definitely more than in other innovation fields of the Delphi '98. Although this criteria cannot be compared directly, it can be stated that the German experts give more medium weight to more financial support. Another reason seems to be that the German experts – although experts - just do not know enough about the Japanese approaches.

Whereas there are nearly no calls for better education as a measure to improve the situation in Germany, the Japanese experts ask for this measure quite often and definitely more often than the Germans. And – also unsurprisingly – the German experts expect more environmental follow-up problems. They do not dramatise, most of the estimations are not dramatically higher, but especially for *microwave sensors which are installed on satellites and capable of defining biomass with an exactitude of less than 1 kg/m2, are in practical use (today: Shuttle SAR-C 1.4 kg); a satellite-assisted, high resolution spectrometrical system in the wave length range from distant infrared to the sub-millimeter area is developed to observe the upper layers of the earth's atmosphere; a system to chart sea and land areas on a global scale is utilised, which works on the basis of a satellite-assisted composite open multi-frequency/ multi-wave radar* and *by using a data collection and storage (DCS) system employing satellites and measuring buoys that ensures long-term, complete and simultaneous data collection, performance analyses and communication over large areas, the service network for information on the fishing and the state of the seas are completed,* the gaps are the highest in the field. The German experts also expect more security problems.

Table 7.12-1: Differences in the estimation of the importance for the economy and the enhancement of human knowledge (in per cent)

Space	Germany		Japan	
	Importance for			
	Enlargement of human knowledge	Economy	Economy	Enlargement of human knowledge
By using a data collection and storage (DCS) system employing satellites and measuring buoys that ensures long-term, complete and simultaneous data collection, performance analyses and communication over large areas, the service network for information on fishing and the state of the seas will be completed.	27,8	87,0	36,5	5,9
Movements in the earth's crust will be measured to within less than 1 cm of displacement by very long baseline interferometers (VLBI), satellite-assisted laser radar, inversion laser distance measurement or LIDAR with synthetic apertures so that the the prediction of earthquakes can be improved.	80,8	59,6	12	19
Following the tethered satellite method, a satellite will be tethered to a space station and used to alter the conditions of gravity, produce electricity and accelerate the payload.	56,3	75,0	53,8	47,6
A system to chart sea and land areas on a global scale is utilized, which works on the basis of a satellite-assisted composite open multi-frequency/ multi-wave radar.	50,0	70,0	26,6	12,8
A satellite-assisted Doppler radar, which makes possible a 3-D definition of wind behaviour within a 500 km observation path from the satellite course, is developed.	44,7	66,0	10,7	22,7

Microwave sensors which are installed on satellites and capable of defining biomass with an exactitude of less than 1 kg/m2, are in practical use (today: Shuttle SAR-C 1.4 kg).	32,6	60,9	13,9	11,5
By means of multi-directional radio technology, a satellite-assisted local transmitting system for every region in Germany (e.g. according to federal states) is developed.	2,6	73,7	87,4	3,2
A permanently manned station is built on the moon to carry out activities like soil analyses, scientific observations from the moon or the development of technology to utilize the moon's resources.	79,4	36,5	55,7	79,2

7.13   Big Science Experiments

In the innovation field of the Big Science Experiments, 10 topics can be compared. Whereas in seven cases, the expertise does not differ very much, there is a higher expertise for *drilling technologies* claimed in Japan: *a drilling technology is in practical use which can drill to a depth of 15 km; the general application of radar technology in drilling earth (use inboring rods or drilling head) can be utilised to determine the behaviour of water in the earth's crust* and *equipment to research extreme environments for the area of low gravitational forces, which can support less than 10-6 G for several days, are realised.*

It is also striking that for nine out of the ten topics the importance for the economy and the enlargement of human knowledge differ significantly (table 7.13-1). The German basic research experts regard the topics like in the average of all topics as much more important for the economy than their Japanese counterparts do.

There are also significant differences in the estimation of the time of realisation (table 7.13-2). Three times, the German experts are more optimistic. These are some of the largest differences in the whole Delphi report.

In most cases, there is consensus that the USA is the leading country in the world. But interestingly for the drilling techniques, the German experts claim to be the most advanced in the world. As the expertise for exactly these topics is rather low in Germany, this can be an exaggeration and has to be checked in detail. On the other hand, the Japanese claim to be more advanced for the following three topics: *neutrino detectors are installed following international agreements in many places to*

Table 7.13-1: Differences in the assessment of the importance for the economy and the enlargement of human knowledge (in per cent, see chapter 2.5, grey boxes stress the higher percentage)

Big Science Experiments	Germany		Japan	
	Importance for			
	Knowledge enhancement	Economy	Economy	Knowledge enhancement
A technology will be developed to drill through the earth's crust from the bottom of the sea to the earth's mantle to remove materials.	86,1	66,7	8,5	63,4
Drilling technology will be in use to drill to a depth of 15 km.	84,4	60,0	15,6	50,6
With the help of electron or positron storage rings with a radiation emission under 1 m pGy a highly efficient radiation optics for analyses of radiating installations, e.g. substances under the earth is applied.	56,3	62,5	12,5	87,5
Measuring stations integrating different instruments like seismometers, gradiometers as well as appliances to measure faults, will be set up in drilling holes throughout the country. They can be used for earthquake prediction.	54,9	64,7	12,2	24,5
The general use of radar techology in earth drilling (use in drill or drilling heads) can be utilized to determine the behaviour of water in the earth's crust.	54,8	71,0	16,7	33,3
Large ultra-high vacuum experimental facilities in which WAKE shields (covers to ward off space ions or molecules in the direction of the orbits) are in use.	67,9	75,0	39	78
Precise atomic frequency standard equipment with a stability of 10^{-15} - 10^{-17} are in practical application.	59,6	74,5	59,2	40,8
Neutrino detectors will be installed following international agreements in many places to examine the structure of the earth's interior.	86,0	24,0	4,7	69,8
Equipment to research extreme environments for the area of low gravitational forces, which can support less than 10-6 G for several days, are realised.	81,1	45,9	63,6	69,3

examine the structure of the earth's interior; precise atomic frequency standard equipment with a stability of 10^{-15} - 10^{-17} are in practical application and *equipment to research extreme environments for the area of low gravitational forces, which can support less than 10^{-6} G for several days, are realised.* In these cases, the German experts do not think they are leading at all, but the USA. In the last mentioned case, 0 % of the Germans think the Japanese are the most advanced, the Japanese give themselves a medium assessment but agree insofar as they also think the USA have the highest R&D level.

Unsurprisingly, the Japanese experts call for better education and also unsurprisingly, the German experts expect more environmental problems in general. But there is one exception: regarding the *positron microscope,* the German experts do not expect environmental or security problems at all (therefore: other problems, which are unknown in detail). But the Japanese expect some environmental problems, even (very few) security problems.

Table 7.13-2: Differences in the time of realisation (the median as the base for comparison is marked in grey)

| Big Science Experiments | Germany | | | | Japan | | | |
| | time of realisation | | | | time of realisation | | | |
	Q1	Median	Q2	never (in %)	Q1	Median	Q2	never (in %)
Neutrino detectors will be installed following international agreements in many places to examine the structure of the earth's interior.	2013	2019	2025	10,9	2008	2013	2016	4,7
High-efficiency geometrical optics for analyses of radiating equipment, e. g. substances un-der the earth, using electron or positron storage rings with a radiation emittance under 1 m pGy are in use.	2011	2015	2023	6,6	2018	2024	after 2025	6,3
A positron microscope will be manufactured.	2010	2014	2019	16,7	2016	2019	after 2025	5,9
The general use of radar techology in earth drilling (use in boring rods or drilling heads) can be utilized to determine the behaviour of water in the earth's crust.	2004	2007	2011	12,9	2008	2012	2018	2,8

To sum up this chapter, it can be stated that the differences between German and Japanese estimations are smaller than expected before the Delphi study started. Nevertheless, interesting, and often unexplicable differences can be observed that are worth being examined in more detail. This demonstrates one of the objectives of foresight: to identify differences, problems, questions which should be examined in more detail to prepare for the future.

8   Selected Methodological Problems

The renewed emphasis on government or national (science) and technology forecasting which can be observed around the world is put into perspectives of methodological developments over half this century. This chapter starts with looking back over 50 years of mixed experiences in government or national technology forecasting, as a part of which also the term "foresight" replaced the well-established "forecasting". While the methodological tool-kit changed from mathematical models to more qualitative scenarios or visions, the application of the Delphi method remained the backbone of many foresight activities. In general, the foresight processes can be compared in terms of their comprehensiveness, their science versus industry orientation, and their analytic versus action-oriented targets.

It does not seem premature to discern several new foresight paradigms. From the perspectives of sociology and political sciences, foresight elements seem to be the means of communication (or the "wiring up", see Martin/ Johnston 1998) for the negotiating systems of society. From an economics and management point of view, foresight is helpful for benchmarking and for initiating feedback processes between future demand and present day investment in research and development. From a cultural point of view, the resurrection of foresight in the 1990s seems to be related to growing globalisation and at the same time, the formation of national or regional innovation systems. Finally, in terms of international affairs, supranational foresight seems to become a new venture.

8.1   The application of Delphi '98 results

In many countries, a strategic approach of using foresight data in an implementation phase in order to close the foresight cycle is pursued. In most countries, this "strategy" is not very successful, but nevertheless, the data are used. The German Delphi '98 is an interesting example in this respect.

When the Delphi '98 survey was finished, no strategic plans for implementation existed, although this was regarded as necessary. Money for "marketing" was not available. From the political side, it was mentioned that this "practical application" should be planned. Unfortunately, the report was published in pre-election times for the German Bundestag (parliament). This meant that political exploitation of such studies was extremely dangerous as many of the current problems and their effects always become obvious when looking into the future (e.g. the lack of personnel in the IT area, demographic consequences, problems in the health sector, the consequences of the unpopularity of expensive big science projects especially in the nuclear power field, and many others).

The election passed, and the new government this time was really a new one – representing different parties, persons, and opinions. It took time until the new BMBF was able to decide on follow-up activities in foresight. It was decided to start a new foresight process called FUTUR (the Latin word for "future"), which commenced on June 14, 1999 at a conference "Forward Thinking" in Hamburg. FUTUR was supposed to make use of Delphi data, analyse them, discuss, develop new questions, topics etc. and with this, prepare a new survey. But it took time to organise the process (Cuhls 2000), and when this book was written, no outcomes were perceived and the whole process was being re-structured (see below).

Nevertheless, the work on Delphi '98 was not in vain. There were many different users of the data. Many are even unknown, as they work more or less anonymously. It is not known who made what use of the two volume edition (around 10,000 sold), the downloads from the Internet and the eight newsletters that dealt with special topics, themes or fields and were provided by BMBF to all interested persons, especially in schools.

The following is known to the German Delphi team (see fig. 8.1-1):

The major users of the Delphi '98 were *companies*. Most of them selected those topics and fields in which they were active. They analysed these topics in detail and used the information for different strategic planning purposes, often with a more long-term perspective than usually applied. Some had working groups to analyse and discuss the data or even made further Mini-Delphi studies to gain more in-depth knowledge of the field. Others developed their own strategic high technology lists. For most of them, it was very interesting to know how others (companies or experts from the different kinds of institutions) rate their field. Do all agree? Or is there no consensus? Did the own company overlook certain problems? Where is the conflict potential?

It was especially interesting for companies to know about fields at the borders of their own activities. They know their own fields pretty well, but what if other products coming from different areas replace their technologies? What happens if interdisciplinary research is conducted? What about technology combinations or fusions, or even the combination of production and services? What other frame conditions will change for the companies? For companies, these are the most interesting questions to be answered.

Figure 8.1-1:   Users of the Delphi '98 in Germany

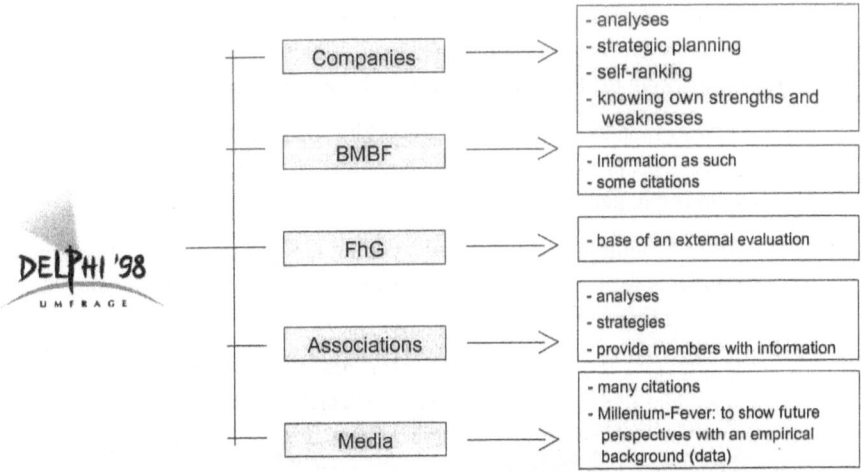

Research institutions and *associations* used the data similarly: for priority-setting, orientation and strategic planning. They also developed high technology lists for themselves, which were added to the traditional world market indicators. The associations sometimes provided their members with the results of their activities.

A very interesting approach was followed for the *system evaluation of the Fraunhofer Society (FhG)* itself. In 1998, the whole society was evaluated on behalf of the German Federal Ministry for Education and Research (BMBF) by an international independent committee (see also Behlau 1998). The questions they followed in their terms of reference were the following:

- Which technology-oriented markets have the greatest dynamics for the German economy?

- Which technology will the market dynamics be based on?

- Does the technological portfolio of the Fraunhofer Society match the developments that are expected, now?

Supported by a scientific secretary from ISI, the Delphi '98 data were used in a unique way to give partial answers to the questions. At first, a new index based on the Delphi criteria *necessity for the improvement of the R&D system, time of realisation*, and *importance of the solutions for the economy* was calculated. The index was applied to the Delphi data in all fields except "Space" and "Big Science". The reason for excluding these fields was that the Fraunhofer Society works in applied sciences linked to markets. The basic research fields especially do not match the criteria of the general Fraunhofer targets. Therefore, only 1,019 of all Delphi topics were included.

According to this, the relevancy was ranked for every thematic field.

The Fraunhofer research fields were defined by persons in the headquarter in Munich who have an overview of the thematic fields the 47 different Fraunhofer institutes are working in. There were four weighting factors:

1. At least two Fraunhofer institutes are working with at least 10 researchers full-time per year on projects for the development of technologies or components that help to implement the Delphi topic.

2. At least one Fraunhofer institute works with at least 10 researchers full-time per year on projects for the development of technologies or components that help to implement the Delphi topic.

3. At least one (2) Fraunhofer institute(s) has (have) a significant research competence and a potential of at least 10 researchers full-time per year to work directly on the realisation of the topic in case there is a concrete project (e.g. topics with a long time horizon).

4. Currently no concrete developments or competencies on a significant scale (more than 10 researchers full-time per year) in the Fraunhofer Society.

Then, a relevancy estimation for Fraunhofer was made. Basis for this is at first the percentage of answers concerning the need for *R&D infrastructure* (between 0 and 100 %). The result is weighted by the time expectancy. The longer the time horizon, the lower the Fraunhofer relevancy.

The R&D needs estimate is divided by the powers of 1.1^x, according to the mean anticipated time horizon. The exponent ^{x} increases with increasing time ^{x} = (mean time – 2000). For the expected year of realisation 2007 therefore the Fraunhofer relevancy is halved compared with the expectation of realisation in 2000. This weighted result corresponds to an index for the present *research intensity* of a Delphi topic.

This index is evaluated again, with the criterion necessary for the Fraunhofer Society of the importance of the innovations for the economic development, that means the growth and the market dynamics which result from the topics.

On average in *all answers* to the Delphi survey, the importance for economic development reached approx. 60 %. As a result, all the data which were under 40 % were undervalued, all the data over 80 % were overvalued. The index of *research intensity* was multiplied by 0.85 for those topics for which the economic development lies under 40 %, and for those which are over 80 %, by 1.15. The remaining topics retain their orginal value of research intensity.

As a result we now have an index for *application-oriented research* (RETIED, for Research, Time, Economic Development), which allows a conclusion on the present Fraunhofer relevancy regarding working on the topic.

Expressed in formulae, the RETIED index would look as follows:

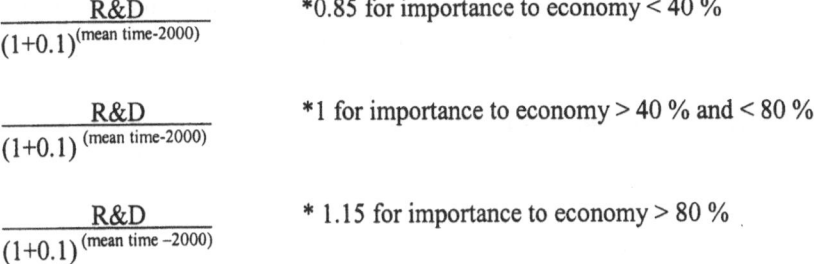

$$\frac{R\&D}{(1+0.1)^{(\text{mean time-2000})}} \qquad *0.85 \text{ for importance to economy} < 40\%$$

$$\frac{R\&D}{(1+0.1)^{(\text{mean time-2000})}} \qquad *1 \text{ for importance to economy} > 40\% \text{ and} < 80\%$$

$$\frac{R\&D}{(1+0.1)^{(\text{mean time}-2000)}} \qquad *1.15 \text{ for importance to economy} > 80\%$$

The topics were sorted for each innovation field according to the RETIED index in descending order and depicted with the appropriate Fraunhofer competence classification. The graphical illustration results in a distribution of the Fraunhofer competencies depending on the RETIED index, which permits a discussion of the quantitative participation of the Fraunhofer Society in the respective future markets.

The aggregated diagrammes make it possible to estimate in which theme areas the Fraunhofer Society is positioned today and with which potential.

Figure 8.1-2:   Fraunhofer evaluation – Results in the field of Information and Communication

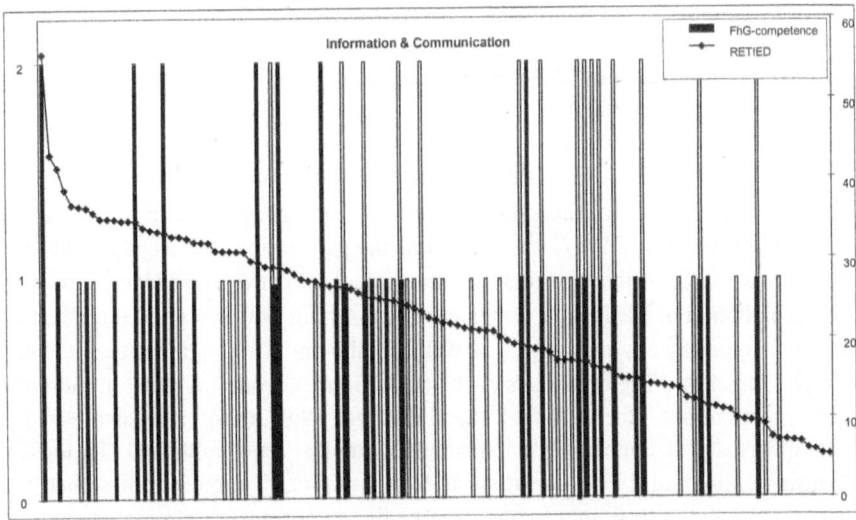

In figure 8.1-2, it can be seen that the field of information technology is represented in the Fraunhofer institutes, but very selectively. The missing fields belong to the most promising markets of the future (e.g. new media). Therefore, the evaluation concluded that there are gaps in information technology in the Fraunhofer Society in general. This was used as *one* of the arguments for the planned fusion of the Fraunhofer Society with another large German research group (press release from 29/9/1999), the German National Research Centre for Information Technology (GMD). Of course, there were more reasons for this decision.

Figure 8.1-3:   Fraunhofer evaluation – Results in the field of Health and Life Processes

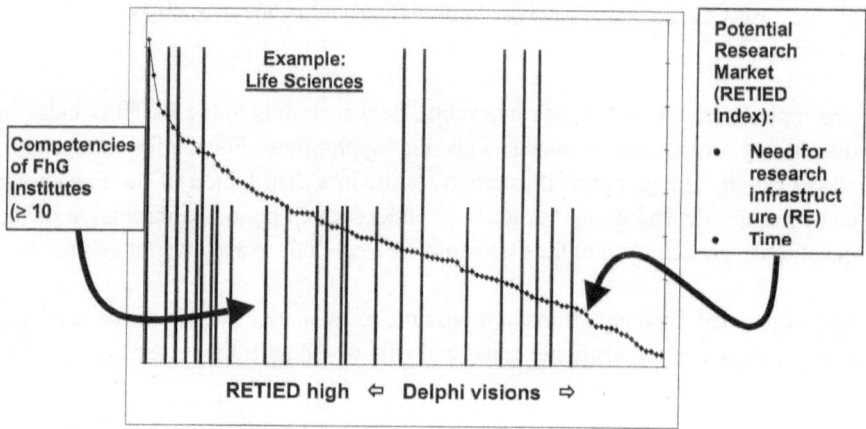

Figure 8.1-3 shows the results concerning Life Sciences. In this field, there were many Delphi topics that related more to longer-term basic research. That means that Fraunhofer is not that much involved here. But for the longer-term view, the evaluation recommended that the FhG needs to keep an eye on this field and become more involved in it.

Another major user of the Delphi results were the *media*. There were two press releases with Federal Minister Jürgen Rüttgers, the first one concerned the end of the field phase in July 1997, the second one the publication of the Delphi '98 report in February 1998. Although there were no explicit measures to involve the media in the exploitation of the Delphi results, the approaching year 2000 had the effect that many newspapers and magazines, as well as radio and TV programmes, mentioned Delphi somehow. The organisers of the study from ISI were involved in more than 120 presentations (for some of them see ISI 1999/2000), and also acted as multipliers for the media. More than 260 articles were published. In the first German Delphi in 1993, the press at first just picked out the topics that were regarded as the critical ones by the experts, like the baby-sitting robot or the 1,000 m high skyscrapers. This time, as there is not much "scientific" data available about the future, the journalists differentiated very much according to their specific public.

Therefore, the interest of the media even grew with time. Even two years later, there were still requests for participation in articles, TV or radio programmes. All 10,000 samples of the report were sold by ISI. As the report is also available on the Internet (www.isi.fhg.de), it is not possible to count how many copies of the report or the short summary have been downloaded. There is still (at the beginning of 2001) at least one telephone call a day at ISI asking when the new report will be published.

This illustrates the demand for foresight data in general, although a more strategic use of the data would have been possible.

8.2 New foresight paradigms in comparing recent national activities in foresight?

Contemporary technology policy has moved away from the inappropriate idea that the state can direct technological developments right down to individual national innovations. Equally outmoded is the idea that the state should be satisfied with the role of a subsidiary supporter of research, leaving the future control of technology to anonymous market processes. Technology policy for the start of the 21st century requires a middle course, i. e. one in which the state plays an active role as an intermediary between negotiating social systems (companies, associations, interest groups, science, consumers, media, representatives of "employers" and "employees", etc.). This intermediary role also must take account of the fact that national technology policy is increasingly restricted in its scope, both from above and below. This is because of the activities of the European Union and the efforts of regional bodies such as the Federal Länder in Germany to promote research on a regional basis (see also the European network FOREN).

The state's new role as active moderator (Kuhlmann 1998, Kuhlmann et al. 1999) necessitates a policy process which is co-ordinated with industry, science and society. However, co-operation does not occur by itself, since too many divergent interests predominate. If there is to be agreement over the possibly selective eligibility for support of technology, dialogues with other social players must be initiated and pursued on a permanent basis. Otherwise, it cannot be expected that lasting co-operation can be achieved or that the platforms to be created for a subject-specific understanding will become more than simply forums for the exchange of information. Don't we need integrated technology foresight to provide the knowledge base for these platforms?

Care has to be taken so that these social negotiations on technological wishes should not stray too far from what is reliably known, and wander into the realms of speculation. In view of the typical recursive phases of science-related technological innovations, it can be generally assumed that everything that will dominate technology impacts in 10 years' time is already recognisable today. However, strategic planning in enterprises is necessary, aiming towards horizons even further in the future, because new technologies - especially those which will contribute to long-sought solutions to problems - must be identified at an early stage.

As far as enterprises are concerned, a considerable improvement of the intramural knowledge base through participation in foresight surveys is reported. There is sporadic evidence that in some companies, during participation in the Delphi, it was felt that too little effort is dedicated towards strategic innovation management and some remedies have been taken (see 8.1). Some companies engaged in own investigations in the direction of an intramural breakdown of the overall national studies towards the special interest of their business areas or establishments, both in the manufacturing and the service sectors. One large chemical company in Germany, especially, started with topics of the Delphi '93 survey, made their own evaluation of the topics and built up a strategy until 2010. In working groups, the information was discussed and distributed. Some smaller-scale comparisons of the business portfolios to the future-oriented areas are also being done in other companies, sometimes assisted by external consultants. These activities are largely confidential.

It is not an easy task to compare the recent national activities in (science and) technology discussed in this issue. Although we think certain general trends in foresight over the past 50 years can be discerned, which was studied in the first chapter for one of the foresight methods only, there is a lot of difference between the countries and within the countries. These final discussion points should be considered as a more impressionistic assessment of the state of the art achieved.

In the category comprehensiveness vis-à-vis selectiveness of the foresight approach one gets the impression that countries with a larger national innovation system tend to be comprehensive and countries with a developing innovation system as well as industrialised countries with a smaller S&T system tend to be more selective. Comprehensive Delphi surveys on a national level have been done in large OECD countries, but also in South Korea. In some developing countries the technique was reduced to certain areas of investigation, and in countries like the Netherlands, selected areas of foresight were agreed upon from the very beginning. Nevertheless, the less comprehensive approaches using *critical technology lists* are also observed in France, Germany, Japan and the United States, and these are larger countries in terms of S&T.

In the category science vis-à-vis industry, it is clear that the major part of foresight is dedicated towards technology or even realised industrial innovation, and only the smaller parts relate to scientific topics. In addition, in some countries, participants from industry were more enthusiastic or more numerous than academics. This is certainly the case in the United Kingdom, but also in Germany, Italy and the Netherlands. In many threshold or developing countries basic research is not well developed. On the other hand, in terms of participation and interest in foresight processes, sometimes staffing from industry was considered very difficult so that a certain gap between orientation of the surveys and participation from the most relevant sectors is reported.

Some national foresight activities are more analytic in nature and others were clearly action-oriented. Among the more analytical approaches we mention the French Delphi survey, while the best case study for a clearly action-oriented foresight programme is the United Kingdom. Other countries navigate between the two poles or start with analysing the problems and then, based on already existing evidence, approach the platforms to bring stakeholders together.

Clearly, there is a reasonable amount of cross-border and comparative work. Here we mention the bilateral co-operations between Japan and Germany or between Germany and France, as well as the orientation of the Korean foresight project towards the Japanese example. Another such case is the Italian study of critical technologies which was deliberately modelled on the German approach, using relevance trees. The repeated surveys in Japan and in Germany provided interesting material to study time series developments and assessments and to put the findings in a historical perspective. Supranational organisations like the OECD and the European Commission invested labour in getting cross-country comparisons and an exchange of experiences (European Commission 1997; OECD 1996, 28).

Against this background of evidence, do we witness new foresight paradigms towards the end of the century? First of all, the bilateral and supranational activities are an indicator that long-term development of science and technology is an international affair. This is not only true for global problems such as climate, but it also reflects the trend to globalisation in R&D. Large multinational enterprises experience certain amounts of mobility with their R&D establishments and their acquisition of human capital, so that countries compete for establishments and try to offer centres of excellence and core competence. Whereas the early forecasting activities were dedicated towards the future of civil R&D of the United States or towards Japan catching up with other countries, now an international or intercultural orientation of foresight is more clearly the case. Some authors note that the international dimension of foresight has some language bias because information relevant for foresight spreads out more quickly within the same language group than between different language clusters. Such observations are reported from German-speaking countries in Central Europe and also Anglo-Saxon countries of the former British Commonwealth.

From an economic perspective, modern foresight has much more to do with benchmarking and feedback processes of economic agents than with systems analysis, cybernetics or operations research as in the early years. Government activities aimed at assessing technical development and providing opportunities for informal technical exchange try to stimulate communication in the so-called "communities", that is, in the informal gatherings of scientists, engineers and business people, where information tends to be exchanged on a non-remunerative basis. Positive external effects occurring within the innovation system can thus be utilised by a company for its own benefit. The deliberate promotion of knowledge flows within these informal circles can thus make an important input into the innovation event. Differences between national economies, relevant for economic growth and competitiveness, arise inasmuch as the corresponding informal circles still tend to be organised nationally, and individual states have different perceptions of what is entailed in identifying future growth areas of technology and in providing opportunities for strategic dialogue to encourage informal panels to exchange ideas.

One serious problem is the acquisition of information from such technology foresight. This sort of inquiry essentially gathers subjective opinions even if the respondents are scientifically trained experts. Even the Delphi process ultimately does not lead to "true" information about the future (which no enterprise can achieve because the future is itself shaped by innovative processes) but, evidence suggests, a reliable database. Furthermore, the feedback training causes the researchers themselves to think seriously about long-term development trends and to clear their thoughts in trans-disciplinary discussions (Grupp 1998, 198). Corporate knowledge procurement cost indicators, however, cannot be deduced from existing foresight investigations of this kind although structural information about the direction of scientific and technical advance can be gleaned and this in turn can be compared, like technology portfolio analysis with the current innovation profile of a company. Here too, there is tremendous scope for additional hitherto untapped resources for sector development.

It was already mentioned that modern polycentric societies can be modelled as negotiating systems from the perspective of sociology. Technology foresight has a potentially important role to play in relation to strengthening these systems in terms of the capacity to learn and innovate. The transition towards the knowledge economy and the increasing importance of institutions of all sorts involved in S&T, requires institutions which form partnerships and strategic alliances and in so doing they need to exchange information. From this a new rationale for (science and) technology foresight arises, which centres on its role in "wiring up" and thereby strengthening the negotiating system of a nation which enables the nation to shape the future so that it better or more quickly meets the longer-term economic and social needs, respectively. Foresight results provide the code to communication between social actors and bring in elements of moderation if conflicting needs are at stake. In this respect the modern use of foresight displays quite some similarity to the real intention of the Delphic oracle in prehistory.

8.3 Choice of experts

Looking at the users of foresight, *the general public* was only informed by the media or presentations about the project until now. Foresight as such remained the task of experts. But there were of course interested persons who asked for a better participation in foresight approaches and who are ready to contribute to the shaping of the future. Many of these people doubt that experts really know what "society" in fact needs or wants. In technology foresight, this means especially a shift from the "technology push" approaches to the more demand-driven "technology pull". But who is "society"? The problem remains to define who is able to participate in such processes as representative, who has enough knowledge to understand topics or questions and who is able to judge in a more objective way than just by subjective "emotions" (Cuhls 2000).

This leads immediately to the question: Who is an "expert"? Who can participate in foresight activities? Can "non-experts" also contribute to this discussion?

Many foresight studies up till now were based on Delphi studies or panels (expert groups). Both concepts involve only "experts". But looking at these studies in more detail, non-experts were involved, too, as the participants also answered questions in fields where they had no particular expertise. In the Japanese and German Delphi studies, among others, for example the respondents were asked for a self-estimation of expertise, which was defined according to the criteria mentioned in chapter 1 (see also Cuhls/ Kuwahara 1994):

As the self-estimation was not regarded as sufficient, in the British Delphi study a co-nomination process was conducted to select the experts (Nedeva et al. 1996) by making the colleagues in the field recommend the experts. The problem here is that the scientific communities are very self-contained. Newcomers or "outsiders" with an uncommon opinion are rarely recommended. The idea of Kuusi (1999), to make detailed interviews with those who claim to be experts is a very interesting idea, but not practicable in large projects.

In the German Delphi studies, the major source of addresses were public databases, publications, trade fairs, and associations, plus many other different recommendations. Sets of related criteria were worked out (e.g. belonging to a certain field of science and technology, regarded to be an expert, involved in research and development or something similar, the younger the better, to increase the number of female persons in the study). It was checked by telephone if the persons really met the criteria.

But this is not enough to secure the good quality of the sample. Also, the carefully selected experts cannot be "high level" experts for a whole technological field. In the Japanese-German comparison, it was also found that the Japanese experts sometimes claimed a higher expertise than the Germans. A reason for this can be a "better" pre-selection of experts in Japan. In Japan, in order to increase the response rate, a postcard was also sent in advance to ask if the person would like to participate. But when some of the German answers were checked in detail and by name, it was observed that German experts systematically understated their own knowledge, because they regarded themselves as experts for the present, but doubted their expertise concerning "the future". In other fields (like Mobility and Transport or Construction and Living), the Japanese expertise varied similarly to the German.

As it is known from "user" feedback, in the German foresight activities, the need is felt to involve more laypersons. The participation of all stakeholders in the innovation system is regarded as necessary to find out if there is consensus or conflict in the different future fields, because the different actors regard different aspects as more important than others or regard the matter from different angles. As one objective of foresight is the combination of different kinds of knowledge by the co-operation of different actors, it was also desired to establish open Internet pages with the possibility of giving comments.

Additionally, the "objective" estimations from scientific studies can be seen more relatively. Subjective wishes of the different groups of actors as well as their career expectations play a very large role – especially for innovations – and should therefore be included in the assessments.

Participation is also relevant for the understanding and acceptance of the results of a foresight activity. All participants should find their views represented somehow. This creates the expectancy that "the rationality as well as the legitimacy of political decisions can be improved" (Hennen 1999, p. 566). It creates a different stimulus to public discussions, especially if these are conducted transparently, so that a sensitisation of the different actors and the public concerning technological, societal, economic, political and other impacts may follow. In this way, foresight processes can also detect in which fields consensus already does or does not exist, so that in the second case a conflict potential can be identified in advance.

These kinds of learning processes were already designed in experiments, e.g. in discursive methods. Discursive approaches make for more rationalised discussions, because they focus on the need to provide arguments. They introduce reasons as a standard of political discussion. Therefore, they correct the strategic (party) intellectuality and argumentative propaganda which is common in the public (mass media) confrontations (van den Daele 1996, 129; also van den Daele 1994, 111ff, and van den Daele 1997). As discursive learning processes, mediation methods are also used in environmental policy (see e.g. WZB-Mitteilungen 1997; examples and conceptional discussions also in Köberle/ Gloede/ Hennen 1997). Discursive approaches should therefore be examined by foresight teams with different methodologies.

These new approaches are also trying to put the general public in the situation of being an expert; consumers especially are important to evaluate innovations. This takes into account that companies listen much more to the demands of their customers, because they sometimes have other wishes than those developed for them by experts in the labs.

The problem here is still the selection of those privileged to participate. In the new German FUTUR process, special attention might be given to all kinds of teachers (in different types of schools and different faculties), for an optimal multiplier effect. As foresight will be relevant for future generations, students in particular will also be asked to participate. A certain knowledge level will nevertheless be a pre-condition.

In a foresight process, the government, or the Ministry for Education and Research (BMBF) in the German case, can implement a policy which relieves the state ("Staatsentlastung") by involving many actors as well as promoting an "active society" in which the importance of a systemic interaction between the relevant actor groups is increasing, as a reaction to deepening interdependent relationships inside the policy arenas (health policy, higher education policy, industrial sector policy) (Messner 1995). Already Etzioni (1969) describes that collective knowledge and therefore the efficient performance of all actors in society and their capability to exchange information result in a steering resource similar to power or money. The state gains power by better legitimation of their decisions following its initiatives for facilitating debates about potential futures.

According to its definition, foresight is oriented towards the decision to go in certain directions. Therefore, it is a special challenge to co-ordinate the different actors in such a way that both, policy-makers and socio-economic actors, are able to perceive an advantage. One of the objectives of foresight is therefore to build upon existing networks or create new ones as lively units and to activate them. These networks cannot be steered directly, but can only be mediated indirectly. One of the connotations here is "Distributed Intelligence" (see fig. 8.3-1) (Kuhlmann et al. 1999): different actors in the networks and the arenas of research and education policy make use of the more or less "intelligent" instruments and information for their strategic future planning. The actor groups have very different value orientations and images of the future. These have to be interrelated in order to facilitate joint strategic use, but not to bring them into the same structure. This can be organised on a joint platform, on which controversial disputes as well as consensus can be sought, but on which the different origins of the concepts still remain visible.

Figure 8.3-1:   Including the different actors of the innovation system

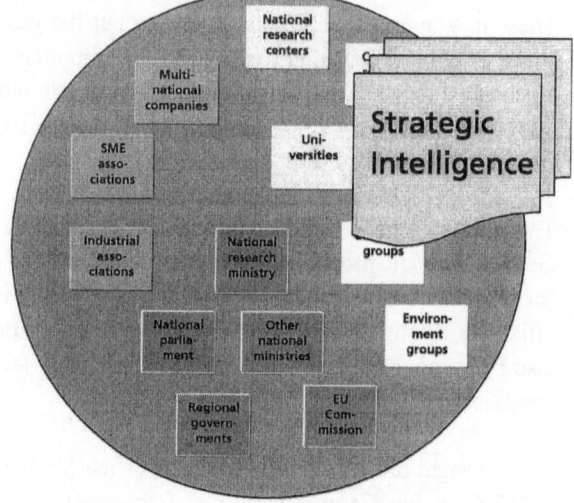

It is very difficult to bring the different actors in the foresight process together. One reason is the differing interests, the others are language and information barriers (e.g. "the public" often does not understand the specific vocabulary of the scientists, or scientists from different disciplines do not understand each other). The identification of the actors (from public databases to member lists of associations, communities, schools) needs very different approaches, which should be followed by neutral parties so that a certain balance is possible. For this, selection criteria should be worked out (e.g. age, expert knowledge, field of expertise, sex, interdisciplinary interests), because in the new processes, not only technical-scientific but also social, societal-cultural, economic, educational and other questions are raised.

Identifying experts and other participants is one task, to link them up and make them communicate is another. Therefore, an interactive network has to be built up. Following the idea of the network, the representatives of the different actor groups will already work out questions in a very detailed manner in different teams, but on the other hand also contact each other independently, without contacting the whole group formally.

Therefore, it is first necessary to find out who the actors in the different fields are (e.g. in the first attempt of the German FUTUR in "Mobility and Communication" or "Health and Quality of Life"). Similar to the co-nomination of the British foresight process, every actor can suggest other persons, so that a dynamic and comprehensive network can be rapidly created. But care must be taken to include also those actors who cannot be found by this method, e.g. outsiders of the communities, persons working along interdisciplinary lines, students, the disadvantaged or those who for other reasons are just not "community members". This can be done to a certain extent by statistical approaches in databases, or by other research with interviews, but it must be done as neutrally as possible. It should be tested if co-nomination can also be used to identify "the least acknowledged person" in the field (anti-co-nomination).

8.4   The integration of existing foresight activities

The integration of different foresight activities is not an easy task. Nearly every approach represents its own concept and methodology. The targets of the approaches are also different. There is no standard solution available as the activities in foresight differ on the following levels:

- national versus international
- different national targets
- sectors or fields
- meso, macro or micro level
- objectives
- selective versus holistic
- (number of) participants
- methodologies used and
- sponsors.

Furthermore, it is only possible to compare themes, topics or whole activities when the activities are conducted at nearly the same time (the time lag should not be more than one year). This is why care is taken to conduct the German Delphi studies in a similar way and at a similar time as the Japanese studies are conducted. Especially quantitative analyses can only be compared under these circumstances. This makes the establishment of foresight databases so complicated. Just to include studies makes no sense if the studies cannot be used in a similar manner.

There might be some possibilities to compare different approaches in a qualitative way. But one has to be very careful not to compare different sectors and different objectives. If participants and their numbers are also different, the bias can be huge.

In this respect, the Futures project conducted at IPTS (IPTS 2000) has to be regarded as an attempt that failed to take these restrictions into account: It compares different studies from different times, e.g. the British Foresight Round 1 from 1993 with the German Delphi '98, although the German Delphi '93 was available as a counterpart from the same time. Fields and methodology were similar, but in the IPTS report there is no hint as to any limitations in the comparison. On the other hand, a comparison of the Austrian Technology (and partly the Society) Delphi with the German Delphi '98 could have been possible as parts of the Austrian Delphi were based on the German one and they were conducted at a similar time, but with different limits and objectives (in the Austrian case, there were only seven technology fields that were oriented towards Austrian needs). The Austrian Delphi was just ignored.

When the reports are regarded in more detail, it can be seen also that different methodologies were mixed in the comparison. And in those sections in which e.g. technologies were compared (technology part), the criteria by which topics for comparison were chosen are not obvious: For instance, in the field of computing (Cahill/ Scapolo 1999, p. 20ff.), Materials and Materials Processing (Cahill/ Scapolo 1999, p. 48ff.), or Transport (Cahill/ Scapolo 1999, pp. 58ff.), only the British, Italian (even using a different methodological approach) and Japanese data are cited, although in the German Delphi '98 the same topics were included but would have signaled different and maybe more recent results. These are just few examples.

The other difficulty concerns the application of the data in general (see chapter 8.1). Here, a more strategic approach already considered in the study design phase should be followed. The ad hoc applications of the Delphi '98 teach us several lessons: there is interest in the application, especially on the side of companies, but as they are not taught how to use the data, the "results" are not fully exploited. On the side of the government, more integration into the discussion of budget planning from the beginning would be very interesting. But this has to be followed very carefully as the notion of *planning* science and technology should not be followed in the sense of "socialist fixed five-year-plans", instead priority-setting for the special concentration on specific sectors or re-orientation is interesting (without neglecting traditional fields or basic research). *Foresight should never be the only source for budget plans.*

It is also necessary to inform media early enough and make them report on what is happening, so that interested actors in the innovation system get early information and thus early access to the data available.

A first step in the integration of foresight studies is to conduct them at the international level, but with a similar design. This cannot be too broad (it becomes too complex), but for example concerning one sector (like the project "AgroFood", Menrad et al. 1998 and 1999). The European Commission can be a facilitator for this.

Another starting point may be the collection of foresight studies in databases that are available, e.g. on the Internet, so that everyone has access to the information. But here also the same criteria apply to the problems mentioned above. The databases must be created in a way that takes the different methodologies, sectors and so on into account .... Even the use of databases as a tool in foresight studies is not as easy as it appears at first sight (see below, the German FUTUR or the British Knowledge Pool are examples of these difficulties).

8.5   New foresight approaches

As the communication effect is stressed in the new foresight processes (Cuhls 2000), many persons and actors from the innovation system are motivated to participate. The new foresight concepts are designed rather as a facilitating factor for communication about the future (awareness raising) than "finding out the truth about the future". Therefore, the definition of expertise is getting much broader. A major target of FUTUR and also the British second foresight activity is to involve more persons than before, to make them communicate, find out *if* there is consensus or not, and create a certain commitment for activities so that they can be planned, started and get support.

The new approaches in foresight can have many objectives (Cuhls 1998)[17]. In the context of policy-making, the most important are to:

- get a larger choice of opportunities, to set priorities and to assess impacts and chances,
- prospect the impacts of current research and technology policy,
- find out new needs, new demands and new possibilities as well as new ideas,
- focus selectively on economic, technological, social and ecological areas as well as to start monitoring and detailed research in these fields,
- elaborate the definition of desirable and undesirable futures and
- start and stimulate continuous discussion processes.

Therefore, a mixture of methods like co-nomination (Nedeva et al. 1996), interviews, available address databases, trade fair lists, publications, conference or

---

17   In some cases, they even have too many at once.

workshop participation in different fields are necessary to identify the participants, as well as the mobilisation of people who are or should be interested (teachers, students, technology transfer personnel, etc.). It is especially necessary to involve the "non-experts" who are interested and motivated in bringing in their - often different - thoughts, especially about the application of the sophisticated ideas scientific experts often have. Therefore, they must be enabled to be or act as an expert, but also to make use of the tacit, implicit or hidden knowledge (Polanyi 1985) of people in general.

The new German foresight activity FUTUR for example is planned to be process-oriented. It takes into account the international experiences and makes use of a mix of methodologies. The process called "FUTUR" (the Latin word for "future"), was first started in 1999 and is now planned to become a more integrated process that no longer separates the different foresight dimensions. Until now, in all countries, the studies centred mainly on scientific-technological questions and influenced the discussion in the other areas. Themes and topics in FUTUR therefore are planned to be broader and include dimensions like education, ethics, social questions, employment and education policy or resource allocation.

The new German foresight process FUTUR does not only look into the future but includes many different stakeholders in the system to find out what things mean for today. It is supposed to draw direct conclusions for technology and research policy, and is therefore directly linked to planning the future budgets. But the decisions on the budget are made by other stakeholders in the system. That means, foresight and FUTUR are not planning, they are the necessary step in identifying future options, making the choices and then acting on them. If there are recommendations, BMBF as the initiator and responsible ministry might use them for planning research programmes. The main aims of FUTUR defined by BMBF were therefore[18]:

- to anticipate future developments and trends in science, research and technology, education, economy, society

- to have a solid base for decisions

- to develop joint long-term visions

- for technologically feasible, ecologically and economically reasonable as well as socially acceptable and need-oriented decisions

- a participative dialogue about the future

- higher acceptability for science and technology.

---

[18]   With changes in FUTUR in 2001, the aims might change and be more focussed.

Figure 8.5-1:   Model of a continuous foresight process

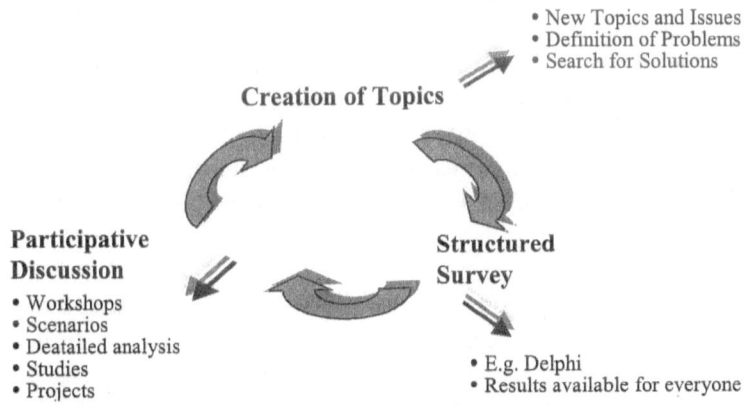

• New Topics and Issues
• Definition of Problems
• Search for Solutions

Creation of Topics

Participative Discussion
• Workshops
• Scenarios
• Deatailed analysis
• Studies
• Projects

Structured Survey

• E.g. Delphi
• Results available for everyone

FUTUR was originally modelled as a continuous foresight process starting with already existing data and knowledge from the German Delphi processes (fig. 8.5-1) and other sources. The results of these previous processes are discussed and restructured. On this basis, new topics, themes, issues are identified, structured and formulated in different working groups so that later on, a new structured survey (maybe a Delphi, maybe something different) can follow. Large parts of the discussion should be conducted via Internet. This was a test if such "virtual workgroups" really do work efficiently. But until now, there is resistance because the participants only had a platform but not the methodology or the issues to work on. They still have to be developed explicitly. Therefore, from 2001 on FUTUR will be re-arranged. There will be physical working groups who work out themes, topics, details, information, assess them, and then describe "pictures of the future" or "scenarios". If these "pictures" are used in a structured survey still has to be decided. But again, these pictures will not be comparable among each other.

Differences in the various foresight approaches result also from the sponsorship and institution operating the studies or processes. In most cases, the research and technology ministry sponsors foresight activities within the framework of a "programme" and it pays external institutions for selective additional studies. Ministries have more power and therefore better access to the different stakeholders in the innovation system. Therefore, the centrality in the organisation of a foresight process seems to be an advantage. But it can also be a disadvantage because of a lack of capacity and experience in the organisation of such large projects. And often "more neutrality" is asked for[19].

---

19   Therefore, the new FUTUR will be organised by a "neutral" consortium outside BMBF.

The new foresight processes also do not only look out for technologies and scientific developments (although these are major driving forces), but look at demographics, society, economics, the environment, politics, culture, ethics and so on. The whole range of themes has widened, thus making foresight more complex and complicated. To introduce a structure into the puzzle, to keep the overview, to select the different parts and paths in more detail and then re-fit them into the whole is a difficult task which is often solved by using a mix of forecasting and foresight methodologies, e.g. scenarios in combination with surveys or Delphi studies, extrapolations in certain fields, simulations or structured interviews with those persons working in the field. These processes can afterwards lead into planning processes.

It is a special challenge to co-ordinate the different actors in such a way that both, policy-makers and socio-economic actors, are able to perceive an advantage. One of the objectives of foresight is therefore to build upon existing networks or create new ones as lively units and to activate them. These networks cannot be steered directly, but can only be mediated indirectly (see 8.3).

In the German FUTUR process, the foresight activities include the collection, the bundling, and the analysis of themes, as well as the diffusion of the results. It is definitely result-oriented. The distinction between a Pre- or Post-Foresight Phase is no longer possible. What is more important is the continuity of the whole foresight process. As the experts stem from different fields, the approaches have been broadened and the experiences applied to other fields (for details see 9.1).

The next challenge is to develop new methodologies for these new approaches to adopt methods from traditional forecasting as well as to build up the link to planning. Some concepts are already available, but the mix of methods will be very useful, depending on the field. Can persons really be made aware of the future tasks instead of remaining in the thought mode of the present (which was often a problem in earlier foresight studies)? How can the tacit or implicit knowledge be activated? Are open, transparent results enough as an incentive or are fixed plans (how and when to do what) necessary in the end?

Another open question is: How can the participants be permanently motivated? Everyone is very busy and has other tasks too. Is the Internet really accepted as the instrument for participation? Or will only "young men with Abitur" (the German high school diploma) participate via the Internet? Are workshops enough as an addition? Or do we need "real events"? How many are necessary? What methods should be used in the groups? Who should organise them? And can the different targets of the different participants really be met? The expectations are very high. Often, the organisers evoked these high expectations. It is already difficult to explain that the concept aims at different targets for different actors. This makes it often very vague. Too vague?

What about the evaluation? The foresight processes have to be evaluated by outsiders (meaning foreigners, in the German case). How can this be organised? The evaluators should understand the language in order to go into details and follow the discussions. And they should have enough knowledge of foresight to judge the matter. What are the targets that can be evaluated when every participant has his own reason to participate? What is the *success* of foresight? If it changes something? If it delivers results for planning? If it motivates the stakeholders in the system to act in a certain way? If foresight supports the communication processes? Or if co-operation partners find each other? How can the new processes be "measured"? And how can the outcome be translated into a certain "strategic thinking" (in the sense of Godet 1997), or even planning? All foresight processes that are going on are still experiments to gather and structure the knowledge about the future on a larger scale. It should be kept in mind that parts of them can also go wrong.

8.6   Databases as a means of foresight

It is often suggested to use Internet and other databases as a tool for foresight purposes. The construction of databases is helpful as a means of information. The problem remains in the target group of the database, the use of it, the updating and therefore the costs to maintain the database as well as the management of the knowledge inherent in it. Just providing the database is no solution. The user must be shown how the information can be turned into knowledge.

With databases, the same problem occurs as with the integration of different foresight activities (see above). The database itself has to be constructed according to the approaches. The same is true for the Internet. The Internet is an interesting media for foresight: as a source of information (similar to a database) as well as a forum for communication. But until now the implementation and application is still quite difficult and in an experimental phase. The samples that can be reached and the persons to judge on future topics or scenarios via the Internet are a biased sample (Kornetzky/ Zoche 1998). One has to be very careful about having control of the sample so that scientific conclusions can be drawn.

If the Internet is used for discussion and assessment purposes in foresight activities, it has to:

- show methodology or a concept
- be target-oriented
- provide pragmatic questions to answer or matters to discuss
- be easily constructed so that everyone can handle it

- be up to date
- be of high quality (here the question remains how to test and control quality)
- be compatible with other existing systems
- be checked if no "dangerous" contents are inherent
- be cheap in construction and maintenance
- be easily restructured and changed
- etc.

By "dangerous" contents, we do not just mean differing opinions, but especially in the German case "Nazi propaganda" might be a danger as well as pornography. The rules must be carefully thought out and also the definition of "dangerous" because censorship is unwanted. And who monitors on the one hand that no censorship occurs, and on the other, that "dangerous" remarks are omitted?

If the Internet is used, a mailer is necessary to inform the participants when something new is happening and if there is new information on the webpage. There is so much information available and there are so many webpages that no one can look at all relevant ones regularly. Therefore, a reminder is convenient.

There should also be incentives for participants to become active. Other media need to inform "the public" or other target groups that Internet pages and information are available. If no one knows about it, no one updates it. The "marketing" here is essential.

8.7　　Why do politics, academia, and business need assessment and foresight?

Today, the architecture for understanding the impacts of a modern science and technology (S&T) policy portfolio is more complicated than ever. The interwoven nature of various types of policies and trade-offs any policy portfolio requires touches on different aspects of the entire quality of life issue. The challenges for a particular policy in the arena of S&T originate from increasing environmental, economic and social problems. The aim of those policies is to make the (national, company) innovation system adaptive enough to meet those challenges. Further, in the area of S&T an increase in the interdisciplinary and transdisciplinary subfields is observed (Reger/ Schmoch 1996; Gibbons et. al 1994; Grupp 1993/1994). Basic or even fundamental research gets into closer contact with industrial research and development (R&D), and science-based technology is pervasive in many industries

(Grupp 1992). More and more unstructured information is available through the different sources.

In politics, in particular, there is an increase in the number of actors and interventions. National policies in some countries are pressured by supranational as well as local and regional bodies. In business, but not in government, globalisation is taking place. For both sectors, and in particular the science establishments, public and private finance is getting short. R&D budgets are limited, and any negative effects of reduced spending are not short-term, i.e., no immediate negative impacts arise. Efficient and effective input-output relations are required. The significance of policies other than monetary ones increases and sufficiently explains the renewed emphasis on foresight and assessment.

This is an attempt to focus the "new complexity" between politics, business innovation strategy, scholarship and foresight. As capitalism is inherently myopic, it focuses on the near-term and discounts the long-term (Greider 1997). It favours near-term investments and resists long-term ones such as infrastructures, education, and R&D. These are left to government and it is no accident that it was the military and its associated aerospace industry which took the lead in promoting the development of diverse forecasting tools (Linstone 1999). The reason is readily apparent: in the four decades since World War II (not considering the last decade) the military confronted the combination of rapidly changing technology, long system lead times, and a perceived long-term cold war threat (Linstone 1999). By contrast, short-term thinking has pervaded the rest of our society and is reflected in the discounting of both distant time and distant space (Linstone/ Mitroff 1994). The focus has to be "here and now". And for this, structured information about the future is necessary.

The renewed interest of technology foresight in the civil realm provokes the question whether this is going to change. We are witnessing new foresight paradigms in our societies and culture, which are not only fed by the interest in "the future" at the beginning of a new century but which try to involve more persons in the decisions concerning R&D, or matters of the future in general (participation) as well as more transparency in the preparation of these decisions (Cuhls 2000b).

9   Outlook

To answer the second part of our basic question: We are not fully convinced that our present society has already made optimal use of foresight for its own progress. There is great potential if we look and think ahead. Of course, some people are very sceptical and see no progress in technology and society. For these, foresight is both costly and irrelevant. But they are often the ones who do not notice changes or are afraid of not being able to adapt to changes. Therefore, in their opinion it is better if no changes occur and the status quo is maintained.

If one looks only at the last 25 to 30 years of German history, in 1972 Germany was in the grip of economic recession; in almost all OECD countries the unemployed were standing on the streets. The Social Democratic Chancellor Willy Brandt resigned over the affair of a GDR spy, and the Federal Minister of Economics, Karl Schiller, a professor of economics, felt he could no longer vouch for his Keynesian policy of demand. On some Sundays at that time you could go for a walk along the motorway, as Chancellor Schmidt had decreed "car-free Sundays" to save petrol. Germany in the 70s did not have: a Federal Ministry for the Environment, a Green Party, relatively well-developed local public transport, separated refuse collection, private TV channels, cable TV, Internet, text programmes, 16 German states, chlorine-free or environmentally friendly paper, industrial robots, genetically engineered tomatoes, CDs, DVDs, videos, milk bottles, a solidarity tax (to help reunified East Germany). There was no State of Bosnia, no German age-care levy; there were no energy-saving lamps! There were no low-energy houses, no PCs in schools, no chipcards ... we could go on with this list. At the beginning of the 70s, two of the authors just started primary school.They still remember times when the things mentioned above were not available, but as they grew up, these items became taken for granted, even necessary. Can you still remember the time you had no standard telephone? And now: maybe you cannot imagine that a few years ago it was no problem to live without a mobile phone – meanwhile, you rely on it.

We would probably be irritated by a return to the year 1970; the changes that have occurred since then, albeit imperceptibly, have been too intensive. However, many problems have remained with us and new problems have cropped up, so that we are sure that foresight, which consists in formulating answers - and re-formulating - complicated questions thrown up at the interface of technology, industry and society, will continue with unabated impetus. In Germany, it is continuing with FUTUR.

9.1   Foresight in Germany: FUTUR's second start

FUTUR was a follow-up process of the Delphi '98. As already mentioned in Chapter 8, it started very ambitiously in 1999.

In the German foresight activities, the need is felt to involve more laypersons. The participation of all stakeholders in the innovation system is regarded as necessary to find out if there is consensus or conflict in the different future fields, because the different actors regard different aspects as more important than others or look at the matter from different angles. As one objective of foresight is the combination of different kinds of knowledge by the co-operation of different actors, it was also desired to establish open Internet pages with the possibility of giving comments.

In the German FUTUR process, the foresight activities include tasks like the collection, the bundling, and the analysis of themes, as well as the diffusion of the results. The activities are supported by 1. selected teams as representatives of different interests, 2. the media, and 3. the Internet to provide information (transparency), as an interactive tool or platform for the discussion. There are also certain milestones of the process which should be followed by evaluations of neutral, if possible, international partners. There will be no Pre- or Post-Foresight-Phase but continuation in the work. As the experts stem from different fields, the approaches have been broadened and the experiences applied to other fields.

At first, topics and questions concerning the future have to be raised. The different kinds of actors will work on these. In this first phase, also new actors in the fields are identified so that a network ("intelligence pool") can be created. The groups meet physically in workshops or virtually via the Internet. The task is to work out questions in discursive processes supported by creativity and other techniques (e.g. scenarios). Also mediated future conferences can be conducted in this phase. The questions must be estimated as worthwhile to be discussed on a broader basis. Possible questions are: Which topics increase the competitiveness of a country? Which fields need kinds of education that were not provided before now? How can this be organised? Which fields harbour a conflict potential? How can the state avoid these conflicts?

When information concerning these topics has been gathered, the topics will be structured according to the questions raised. This will be done in physical working groups. Maybe also large events might take place (similar to open space conferences) in order to involve interested citizens as participants and to have a larger input of expectations.

From these, "pictures of the future" (e.g. living in 2020) are worked out. Some of them will be "real" scenarios, others just descriptive pictures. Scenario methodologies will be used for this. These "pictures" will already be used for further work in informing the public and identifying research areas of the future for the BMBF, as well as the implementation of parts of these pictures.

After so many questions, pictures, and topics have been raised, a special focus can be obtained with a comprehensive, systematic and comparative survey among the most important actor groups. This central study[20] should include the objective estimation of (e.g. technological) potentials as well as the subjective wishes (also of the different parties involved), whose influence on future developments cannot be underestimated. The Delphi methodology with its feedback to all participants is one of the methods that can be used in this context. Scenario workshops with feedback could be another possibility. "Pictures of the future" would be described to illustrate the developments.

When the intermediate results of the structured surveys are available, the implementation and discussion follows. For this, teams from the different actor groups are founded in order to discuss the future topics in a target-oriented manner. These discussions follow applied participative methodologies. "Planungszellen" (planning cells), consensus conferences, focus groups, mediation (especially in the case of conflicts) or local scenario workshops can be used to structure the discussion. These methodologies can only be applied if the topic itself is already clearly stated. The conception, mediation, and monitoring will be designed by a neutral partner. For in-depth discussion in specific fields, a specialist mediator is needed who is able to ask the "right" questions and to follow the debate. This mediator needs to behave as neutrally as possible.

Especially in the implementation phase, it must be determined exactly which targets are to be pursued and whose targets they are. For the Federal German Ministry of Education and Research (BMBF), maybe setting priorities in certain research programmes or working out new research programmes can be a target (Cuhls 2000b). Other targets can be more interesting for industry (e.g. tax reductions for certain fields, financing). Specialist evaluations, the recommendations of the public ("Bürgerempfehlung") for certain projects (e.g. should the German magnetic train Transrapid be built or not?) or definite plans for measures (financing of a specific research institution, the foundation of new institutions or closing down institutions no longer needed) can also be objectives.

From the implementation, new topics and questions evolve, and a new topic generation phase can start. This makes the foresight process a continuous one. The management of the process is performed by three major actors: one part of the tasks is to monitor the process, prepare the concepts and analysis, and organise the

---

20 In the "new FUTUR" approach, the central study will follow later on. At first, only "pictures" are developed.

Figure 9.1-1:   Phases of a foresight process

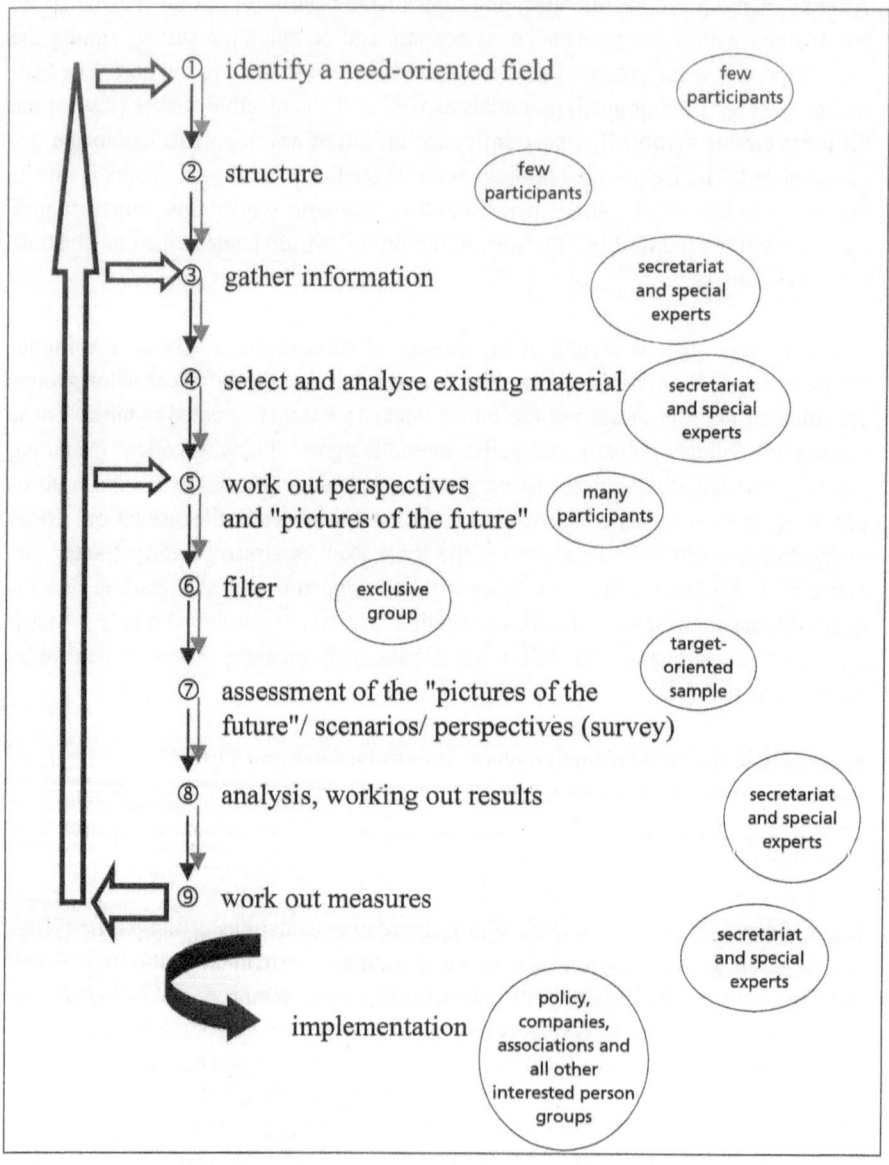

evaluation of the whole process. The Fraunhofer Institute for Systems and Innovation Research (ISI) might play a major role in this part.

The second major part is that of decision-maker and mediator of the whole process, to be played by the BMBF in general. The minister herself will take responsibility for this process. The various media and especially the Internet are the liaison partners in the process. A management and co-ordinating institution is also needed to manage the Internet, invite the different kinds of actors and co-ordinate the different workshops in the process (in FUTUR, this will be done by the Institute for Organisation Communication, IFOK). This ideal concept is already a network that co-operates very closely. The Internet is the platform for transparency and for presenting results and intermediate results. The external presentation part will play a large role because without activating and motivating people (also the public) to participate, it will be difficult to keep the process alive. In the Internet, information will be provided on already existing material as well as links to other homepages that are interesting in this context.

The panels are made up of the different actors. They work dependently and independently – according to their wishes. A person who is well known in the field and who is able to motivate the participants might be one of the key persons in the group. The teams work out their topics using the different foresight and creativity methods. They are assisted by professional mediators and are asked to deliver a defined outcome after a certain time. It can be a report, recommendations, short descriptions of new trends, CD ROMs, workshops, the organisation of a conference, founding of a specific network or task forces. But the outcome is necessary to get some visible return out of it, and it should also be of use for the participants themselves.

The German foresight process FUTUR is re-arranged as the difficulties of co-ordination from the first start are apparent. The network of actors and participants is already large and becoming increasingly larger. Also the organisation is co-ordinated in a team consisting of different institutions. The whole process management is supposed to be as self-learning as the process itself. It also needs time to convince all possible partners, to provide the financial support and solve the technical and organisational problems of an Internet page.

But the major difficulty is to convince actors in the innovation process that they really benefit from participating in foresight studies, i.e. investing their time in the process. If they are convinced, additional, equally competent participants have to be found and persons from very different fields integrated, not forgetting potential partners. In some fields, the interest is already very great – especially in industry. In other fields, this is not the case, so that it is difficult to involve the "right" mixture of participants.

Again, we have to ask: What is the *success* of FUTUR? If it changes something? If it motivates the stakeholders in the system to act in a certain way? If foresight supports the communication processes? Or if BMBF starts new programmes according to their newly developed "Leitvisionen" (leading visions)? How can a process like this be "measured"? And how can the outcome be translated into a certain "strategic thinking" (in the sense of Godet 1997)?

Many questions are still open, and not all of them can be answered yet. This chapter shows some of the limitations foresight processes have to face. But change is necessary to be better prepared for the tasks of the future. In Germany, we would comment this with "Only change is unchanging" (Nur der Wandel ist beständig).

9.2   Challenges of tomorrow

Many challenges of the future remain. Some of them are similar to the challenges of the past. Some of them – like globalisation – are just more challenging, because of the new forms and possibilities they are embedded in.

Technology policy therefore needs and already establishes a new course with a new role of the state. Top-down dictatorship is out – and was never possible in democracies. The new role of the state is rather an intermediary one among the different stakeholders in the social system (companies, associations, interest groups, science, media, consumers, employers, employees, etc.), taking into account the fact that national technology policy is increasingly restricted in its scope, both from above and below. And there is not only the national state as an actor, anymore, but also the European Union or regional bodies like the federal Länder in Germany who are actively promoting their own technology policy. Additionally, the "objective" estimations from scientific studies can be seen more relatively. Subjective wishes of the different groups of actors as well as their career expectations play a very large role – especially for innovations – and should therefore be included in the assessments. Thus, in many countries, the "public acceptance" of innovations is regarded as critical.

Participation is also relevant for the understanding and acceptance of the results of a foresight activity. These new approaches are also trying to put the general public in the situation of being an expert; consumers especially are important to evaluate innovations (see chapter 8.3). This takes into account that companies listen much more to the demands of their customers, because they sometimes have other wishes than those developed for them by experts in the labs.

The problem here is still the selection of those privileged to participate. In the German FUTUR, special attention might be given to all kinds of teachers (in different types of schools and different faculties), for an optimal multiplier effect. As foresight will be relevant for future generations, students in particular will also be asked to participate. A certain knowledge level will nevertheless be a pre-condition.

It is very difficult to bring the different actors in the foresight process together. One reason are the differing interests, the others are language and information barriers (e.g. "the public" often does not understand the specific vocabulary of the scientists, or scientists from different disciplines do not understand each other). The identification of the actors (from public databases to member lists of associations, communities, schools) needs very different approaches, which should be followed by neutral parties so that a certain balance is possible. For this, selection criteria should be worked out (e.g. age, expert knowledge, field of expertise, sex, interdisciplinary interests), because in the new processes, not only technical-scientific but also social, societal-cultural, economic, educational and other questions are raised.

Identifying experts and other participants is the one task, to link them and make them communicate is the other. Therefore, an interactive network has to be built up. Following the idea of the network, the representatives of the different actor groups will already work out questions in a very detailed manner in different teams, but on the other hand also contact each other independently, without contacting the whole group formally.

Therefore, it is first necessary to find out who the actors in the different fields are (e.g. in the first attempt of FUTUR in Mobility and Communication or Health and Quality of Life). Similar to the co-nomination of the British foresight process, every actor can suggest other persons, so that a dynamic and comprehensive network can be rapidly created. But care must be taken to include also those actors who cannot be found by this method, e.g. outsiders of the communities, persons working along interdisciplinary lines, students, the disadvantaged or those who for other reasons are just not "community members". This can be done to a certain extent by statistical approaches in databases, or by other research with interviews, but it must be done as neutrally as possible. It should be tested if co-nomination can also be used to identify "the least acknowledged person" in the field (anti-co-nomination).

One of the objectives of the German foresight process FUTUR is the openness for other topics and themes, plus the participation of non-experts. The interpretation of already existing data and the creation of new questions cannot be independent of different opinions on problems, especially for the assessment of new technology and its application effects. Therefore, the identification of future topics and the selection of those which deserve in-depth treatment needs the participation of those involved and interest groups or lobbies. Furthermore, the specific knowledge of the different actor groups shall be applied in order to detect risks and chances, but also to arrive at innovative solutions for existing problems. The representation of as many opinions or sides of the problem as possible from the different actor groups in the various fields can improve the analytical depth of the assessments and the commitment for a topic. But it also complicates the methodologies for decision-making resulting from this.

The state's new role as active moderator necessitates a policy process which is co-ordinated with industry, science and society. However, co-operation does not occur by itself, since too many divergent interests predominate. If there is to be agreement over the possibly selective eligibility for support of technology, dialogues with other social players must be initiated and pursued on a permanent basis. Otherwise, it cannot be expected that lasting co-operation can be achieved or that the platforms to be created for a subject-specific understanding will become more than simply forums for the exchange of information. Don't we need integrated foresight to provide the knowledge base for these platforms?

Care has to be taken so that these social negotiations on technological wishes should not stray too far from what is reliably known, and wander into the realms of speculation. In view of the typical recursive phases of science-related technological innovations, it can be generally assumed that everything that will dominate technology impacts in 10 years' time is already recognisable today. However, strategic planning in enterprises is necessary, aiming towards horizons even further in the future, because new technologies - especially those which will contribute to long sought solutions to problems - must be identified at an early stage. Some large German companies just started own comprehensive approaches in this context.

As far as enterprises are concerned, a considerable improvement of the intramural knowledge base through participation in foresight surveys is reported. There is sporadic evidence that in some companies, during participation in the Delphi, it was felt that too little effort is dedicated towards strategic innovation management and some remedies have been taken. Some companies engaged in own investigations in the direction of an intramural breakdown of the overall national studies towards the special interest of their business areas or establishments, both in the manufacturing and the service sectors (see also chapter 8.1). But the interest of companies in "more information about the future" is obvious. To gain this remains one of the challenges for foresight.

It is not an easy task to compare the recent national activities in (science and) technology activities discussed in this issue. Although we think one can discern certain general trends in foresight over the past 50 years, which was studied for one of the foresight methods only, there is a lot of difference between the countries and inside the countries. These final discussion points should be considered as a more impressionistic assessment of the state of the art achieved shortly before the century turns. From the methodological point of view, nearly all countries are experimenting with a mix of methods.

Regarding comprehensiveness vis-à-vis selectiveness of the foresight approach, one gets the impression that countries with a larger national innovation system tend to be comprehensive and countries with a developing innovation system as well as industrialied countries with a smaller S&T system tend to be more selective (see above). The objective for foresight is therefore most important. No country should invent the wheel again. In some cases it might be wise to rely on existing material but re-evaluate it for own purposes and objectives. And not every country is able to switch to a high technology country in very few years – just because foresight studies of other countries advise investing in high technologies. Not every country can afford this – and not every country needs it economically or culturally. In some cases, selective studies might be the better choice in large countries also.

In the category science vis-à-vis industry, it is clear that the major part of foresight is dedicated towards technology or even realised industrial innovation, and only the smaller parts relates to scientific topics. In addition, in some countries, participants from industry were more enthusiastic or more numerous than academics. This is certainly the case in the United Kingdom, but also in Germany, Italy and the Netherlands. In many threshold or developing countries basic research is not well developed. On the other hand, in terms of participation and interest in foresighting processes, sometimes staffing from industry was considered very difficult so that a certain gap between orientation of the surveys and participation from the most relevant sectors is reported. But the demand to broaden the scope of topics and participants is obvious.

Some national foresight activities are more analytic in nature and others were clearly action-oriented. Clearly, there is a reasonable amount of cross-border and comparative work. We already mentioned the bilateral co-operations between Japan and Germany, and between Germany and France, as well as the orientation of the Korean foresight project towards the Japanese example as mentioned in chapter 8. We seem to witness new foresight paradigms towards the beginning of the new century. First of all, the bilateral and supranational activities are an indicator that long-term development of science and technology is an international affair. This is not only true for global problems such as the climate problem, but it also reflects the trend to globalisation in R&D. Large multinational enterprises experience certain amounts of mobility with their R&D establishments and their acquisition of human capital so that countries compete for establishments and try to offer centres of excellence and core competence. Whereas the early forecasting activities were dedicated towards the future of civil R&D of the United States or Japan's catching up with other countries, now more clearly an international or intercultural orientation of foresight is the case. Some authors note that the international dimension of foresighting has some language bias because information relevant for foresight spreads out quicker within the same language group than between different language clusters. Such observations are reported from German-speaking countries in Central Europe and also Anglo-Saxon countries of the former British Commonwealth.

From an economic perspective modern foresight has much more to do with bench-marking and feedback processes of economic agents than with systems analysis, cybernetics or operations research based on "forecasting" as in the early years (Cuhls 2000b). Government activities aimed at assessing technical development and providing opportunities for informal technical exchange try to stimulate communication in the so-called "communities", that is, in the informal gatherings of scientists, engineers and business people, where information tends to be exchanged on a non-remunerative basis. Thus, the "5 Cs" described by Martin (1995) are still relevant in foresight.

Positive external effects occurring within the innovation system can thus be utilised by a company for its own benefit. The deliberate promotion of knowledge flows within these informal circles can thus make an important input into the innovation event. Differences between national economies, relevant for economic growth and competitiveness, arise inasmuch as the corresponding informal circles still tend to be organised nationally, and individual states have different perceptions of what is entailed in identifying future growth areas of technology and in providing opportunities for strategic dialogue to encourage informal panels to exchange ideas.

One serious problem is the acquisition of information from such technology foresight. This sort of inquiry essentially gathers subjective opinions, even if the respondents are scientifically trained experts. Even the Delphi process ultimately does not lead to "true" information about the future (which no enterprise can achieve because the future is itself shaped by innovative processes), but evidence suggests, a reliable database. Furthermore, the feedback training causes the researchers themselves to think seriously about long-term development trends and to clarify their thoughts in trans-disciplinary discussions (Grupp 1998, p. 198), as well as in the clear formulation of future issues – to write them down is a step further from just "having in mind". This is a clear step from tacit to explicit knowledge.

Corporate knowledge procurement cost indicators, however, cannot be deduced from existing foresight investigations of this kind, although structural information about the direction of scientific and technical advance can be gleaned and this in turn can be compared like technology portfolio analysis with the current innovation profile of a company. Here too, there is tremendous scope for additional hitherto untapped resources for sector development.

It was already mentioned that modern polycentric societies can be modelled as negotiating systems. The role of foresight in relation to strengthening these systems in terms of the capacity to learn and innovate can be an important one. The transition towards the knowledge economy and the increasing importance of institutions of all sorts involved in S&T, requires institutions which form partnerships and strategic alliances and in so doing they need to exchange information in this arena of players. From this a new rationale for science and technology foresight arises which centres on its role in "wiring up" and thereby strengthening the negotiating system of a nation which enables them to shape the future so that the longer-term economic and social needs can be met better or quicker, respectively. Foresight results provide the code to communication between social actors in the arena and bring in elements of moderation if conflicting needs are at stake. In this respect the modern use of foresight displays quite some similarity to the real intention of the Delphic oracle in prehistory.

And "as only change is unchanging", maybe we will be able to do foresight on a concerted EU or even global level ...

Literature

Bardecki, M. J. (1984): Participant's Response to the Delphi Method: An Attitudinal Perspective, in: Technological Forecasting and Social Change, vol. 25, pp. 281-292.

Bea, F. X.; Dichtl, E. and Schweitzer, M. (1989): Allgemeine Betriebswirtschaftslehre, Band 2: Führung, Fischer, Stuttgart/ New York.

Behlau, H. (1998): Reflexion der Aktivitäten der Fraunhofer-Gesellschaft an den Zukunftsvisionen der Delphi '98 Umfrage, in: Cuhls, K.; Grupp, H. and Blind, K. (eds.): Delphi '98 - Neue Chancen durch strategische Vorausschau, pp. 17-20. Tagungsband der Tagung in der Deutschen Bibliothek in Frankfurt/Main 1998, Karlsruhe.

Blind, K; Cuhls, K; Grupp, H. (1999): Current Foresight Activities in Central Europe, in: Technological Forecasting and Social Change, Special Issue on National Foresight Projects, vol. 60, pp. 15-35.

Brockhaus Encyclopedia (1993), Edition 19, Mannheim, p. 343.

Brockhaus (1999): Die Enzyklopädie, 20th edition, Leipzig, Mannheim 1999.

Bundesministerium für Forschung und Technologie (Federal Ministry for Research and Technology, BMFT, ed.) (1993): Deutscher Delphi-Bericht zur Entwicklung von Wissenschaft und Technik (German Delphi Report on the Development of Science and Technology), Bonn.

Bundesministerium für Wissenschaft und Verkehr (Federal Ministry for Research and Traffic, ed.) (1998): Delphi Report Austria I, II, III. Technologie Delphi – Konzept und Überblick, Wien .

Bush, V. (1945): Science, The Endless Frontier, National Science Foundation, Washington, D.C.

Cahill, E. and Scapolo, F. (eds.) (1999): Technology Map. Futures Report Series 11 of the Institute for Prospective Technologies (IPTS), EUR 19031 EN, Seville.

Catell R. B. (1978): The scientific use of factor analysis in behavioral and life sciences, New York.

Child, D. (1975): The Essentials of Factor Analysis, London.

Coates, J. F. (1985): Foresight in Federal Government Policymaking, in: Futures Research Quarterly, Summer 1985, pp. 29-53

Corbin, S. S. and Chiachiere, F. J. (1995): Validity and Reliability of a Scale Measuring Attitudes toward Foreign Language. in: Educational and Psychological Measurement, vol. 55, 2, 258-267.

Cuhls, K. (1998): Technikvorausschau in Japan, Heidelberg: Physica (Technik, Wirtschaft und Politik 29).

Cuhls, K. (2000a): Opening up Foresight Processes, in: Économies et Sociétés, Série Dynamique technologique et organisation, no. 5, pp. 21-40.

Cuhls, K. (2000b): Wie kann ein Foresight-Prozess in Deutschland organisiert werden? (How can a foresight process in Germany be organised?) Gutachten. Medien- und Technologiepolitik, Friedrich-Ebert-Stiftung, Bonn.

Cuhls, K. and Kuwahara, T. (1994): Outlook for Japanese and German Future Technology. Comparing Technology Forecast Surveys, Heidelberg: Physica (Technology, Innovation, and Policy 1).

Cuhls, K.; Blind, K.; Grupp H. (1998): Delphi '98, Umfrage. Studie zur globalen Entwicklung von Wisenschaft und Technik, Karlsruhe.

Cuhls, K.; Breiner, S. and Grupp, H. (1995): Delphi-Bericht 1995 zur Entwicklung von Wissenschaft und Technik - Mini-Delphi – (Delphi Report on the Development of Science and Technology – Mini-Delphi), Karlsruhe (same as Brochure of BMBF, Bonn 1996).

Dalkey, N. (1969): An Experimental Study of Group Opinion: The Delphi Method, in: Futures, no. 1, pp. 408-420.

Dalkey, N. and Helmer, O. (1963): An Experimental Application of the Delphi Method to the Use of Experts, Management Sciences, no. 9, pp. 458-467.

Dalkey, N. C. (1969): The Delphi Method: An Experimental Study of Group Opinion, typescript, Rand, Santa Monica.

Dreher, C. and Schirrmeister, E. (2000): The long road to closed cycle management. Communications from the Survey on Innovations in Production, PI-Mitteilung, no. 18, FhG-ISI, Karlsruhe.

Etzioni, A. (1968): The Active Society, New York.

European Commission (eds) (1997): Second European Report on S&T Indicators 1997, EUR 17639, European Commission, Luxembourg.

Flechtheim, O. (1968): Futurologie - Möglichkeiten und Grenzen, Frankfurt/M., Berlin.

Forrester, J. W. (1971): Der teuflische Regelkreis. Das Globalmodell der Menschheitskrise, Stuttgart.

Fraunhofer Institute for Systems and Innovation Research (2000): Annual Report 1999/2000, Karlsruhe.

Gibbons, M., Limoges, C., Nowotny, H., Schwartzman, P., Scott, P. and Trow, M. (1994): The New Production of Knowledge, London: Sage publications.

Godet, M. (1997): Scenarios and Strategies. A Toolbox for Problem Solving, Cahiers du LIPS, Special Issue, Paris.

Gordon, T. J. und Helmer, Olaf (1964): Report on a Long-Range Forecasting Study, Rand Corporation, Santa Monica/ California.

Greider, W. (1997): One World, Ready or Not: The Manic Logic of Global Capitalism, New York: Simon & Schuster.

Grupp, H. (1992): Dynamics of Science-Based Innovation, Heidelberg/New York: Springer.

Grupp, H. (ed.) (1993): Technologie am Beginn des 21. Jahrhunderts, Heidelberg: Physica-Verlag (Technik, Wirtschaft und Politik, 3, 2nd edition 1995); abridged English version: Grupp, H.: Technology at the Beginning of the 21st Century, in: Technology Analysis & Strategic Management (1994) no. 6, pp. 379-409.

Grupp, H. (1998): Foundations of the Economics of Innovation - Theory, Measurement and Practice, Cheltenham: Edward Elgar. (German Version: Messung und Erklärung des Technischen Wandels, Berlin, Heidelberg, New York: Springer, 1997).

Grupp, H. (1999) (ed.): Technological Forecasting and Social Change, Special Issue on National Foresight Projects, vol. 60 (1999) no. 1, New York: Elsevier Science.

Häder, M. and Häder, S. (1995): Delphi und Kognitionspsychologie. Ein Zugang zur theoretischen Fundierung der Delphi-Methode. ZUMA-Nachrichten, 37, Mannheim.

Harman, H. H. (1967): Modern factor analysis, 2nd edition, Chicago.

Hayduck, L. A.; Ratner, P. A.; Johnson, Joy L. and Bottorff, J. L. (1995): Attitudes, Ideology, and the Factor Model. in: Political Psychology, vol. 16, no. 3, 479-507.

Helmer, O. (1964): Looking Forward. A Guide to Futures Research, Beverly Hills, London, New Delhi.

Helmer, O. (1966): Social Technology, New York, London: Basic Books, 1966.

Helmer, O. (1967): Analysis of the Future: The Delphi method, Santa Monica: Rand Corporation.

Helmer, O. (1977): Problems in Futures Research - Delphi and Causal Cross-Impact Analysis, in: Futures, no. 9, pp. 17-31.

Helmer, O. and Rescher, N. (1959): On the Epistemology of the Inexact Sciences, in: Management Science, no. 6, pp. 5-52.

Hennen, L. (1999): Partizipation und Technikfolgenabschätzung, in: Bröchler, S.; Simonis, G. and Sundermann, K. (eds.): Handbuch Technikfolgenabschätzung, vol. 2, Berlin: Ed. Sigma.

Hirowatari, S. (2000): Japan's National Universities and Dokuritsu Gyōsei Hōjin-ka, in: Social Science Japan, Newsletter of the Institute of Social Science University of Tokyo, vol. 19, pp. 3-13.

Hounshell, D. A. (1996): The Medium is the Message, or How Context Matters: The RAND Corporation Builds on Economics of Innovation, Pittsburgh, no date, from the author in June 1996.

Hughes, Th. P. (1998): Rescuing Prometheus, New York: Pantheon Books and Toronto: Random House of Canada Limited.

Hurtig, E. and Stiller H. (1984): Erdbeben und Erdbebengefährdung [Earthquake and Danger], Berlin.

IIT Research Institute (1968): Technology in Retrospect and Critical Events in Science (TRACES), National Science Foundation, Washington, D.C.

Institute for Prospective Technological Studies (IPTS, ed.) (2000): The IPTS Futures Projects Synthesis Report, EUR 19038 EN, Seville.

Irvine, J. and Martin, B. R. (1984): Foresight in Science. Picking the Winners, London and Dover: Francis Pinter.

Isenson, R. S. (1967): Technological forecasting lessons from Project HINDSIGHT, Paper presented the Technology and Management Conference, Harvard University, 22 May.

Jantsch, E. (1967): Technological Forecasting in Perspective, OECD, Paris.

JETRO (ed.) (1991): Earthquake Prediction Technology – From AIST's Basic R&D programs -, in: New Technology Japan, vol. 19, no. 3, p. 10.

Kagaku Gijutsuchô (Science and Technology Agency, ed.) (1998): Kenkyû Kaihatsu no Hyôka no Genjô (Current Situation of the Evaluation of Research and Development), Tokyo.

Kagaku Gijutsuchô Keikakukyoku (Science and Technology Agency, Planning Bureau, ed.) (1971): Gijutsu Yosoku Hôkokusho (Report on Technology Forecast Survey), Tokyo.

Kahn, H. and Wiener, A. J. (1967): The Year 2000 [Ihr werdet es erleben, Voraussagen der Wissenschaft bis zum Jahre 2000], Wien, München, Zürich.

Kaiser, H. F. (1974): An index of factorial simplicity. in: Psychometrica, vol. 39, pp. 31-36.

Kaplan, A., Skogstad, A. L. and Girshick, M. A. (1950): The Prediction of Social and Technological Events, in: The Public Opinion Quarterly, XIV, pp. 93-110.

Kecskes, R. and Wolf, Ch. (1993): Christliche Religiosität: Dimensionen, Meßinstrumente, Ergebnisse. in: Kölner Zeitschrift für Soziologie und Sozialpsychologie, vol. 45, pp. 270-287.

Klein, H. (1998): Review of Cuhls, Kerstin and Kuwahara, Terutaka, Outlook for Japanese and German Future Technology: Comparing Technology Forecast Surveys, in: International Journal of Forecasting, vol. 14, no. 2, 301-303.

Köberle, S.; Gloede, F. and Hennen, L. (eds.) (1997): Diskursive Verständigung? Mediation und Partizipation in Technik-Kontroversen, Baden-Baden: Nomos.

Kornetzky, S. and Zoche, P. (1998): Internet User Survey. Ausgewählte Ergebnisse einer Parallelumfrage zur Nutzung der Neuen Medien einschließlich eines Methodenvergleiches, Karlsruhe.

Kuhlmann, S. (1998): Politikmoderation. Evaluationsverfahren in der Forschungs- und Technologiepolitik, Baden-Baden: Nomos.

Kuhlmann, S. et al. (1999): Improving Distributed Intelligence in Complex Innovation Systems, Final Report of the Advanced Science & Technology Policy Planning Network (ASTPP), Karlsruhe.

Kuusi, O. (1999): Expertise in the Future Use of Generic Technologies. Epistemic and Methodological Considerations Concerning Delphi Studies, Helsinki: Helsinki School of Economics and Business Administration.

Linstone, H. A. (1997): The Changing Role of Forecasting, to be published.

Linstone, H. A. (1998): Multiple Perspectives Revisited, IAMOT Conference, Orlando.

Linstone, H. A. (1999): Decision Making for Technology Executives. Using Multiple Perspectives to improve performance, Boston, London: Artech House.

Linstone, H. A. und Simmonds, W. H. C. (1977) (eds.): Futures Research: New Directions, Reading/ Mass.

Linstone, H. A. with Mitroff, I. I. (1994): The Challenge of the 21st Century: Managing Technology and Ourselves in a Shrinking World, Albany: State University of New York Press.

Linstone, H. A. and Turoff, M. (eds) (1975): The Delphi Method – Techniques and Applications, Reading: Addison-Wesley.

Martin, B. R. and Johnston, R. (1999): Technology Foresight for Wiring Up National Innovation Systems: Experiences in Britain, Australia, and New Zealand, in: Grupp, Hariolf (ed.): Technological Forecasting and Social Change, Special Issue on National Foresight Projects, vol. 60, no. 1, pp. 37-54.

Martin, B. R. (1995): Foresight in Science and Technology. in: Technology Analysis & Strategic Management, vol. 7, no. 2, pp. 139-168.

Martino, J. P. (1983): Technological Forecasting for Decision Making, 2nd edition, North Holland, New York, Amsterdam, Oxford.

Martino, J. P. (1993): Technological Forecasting for Decision Making, 3 edition, New York: McGraw-Hill.

Meadows, D. H., Meadows, D. L., Randers, J. and Behrends III, W. W. (1972): The Limits to Growth, New York: Universe Books.

Menrad, K.; Agrafiotis, D.; Enzing, Chr.; Lemkow, L. and Terragni, F. (1999): Future Impacts of Biotechnology on Agriculture, Food Production and Food Processing, Heidelberg: Physica (Technology, Innovation, and Policy 10).

Menrad, K.; Koschatzky, K.; Maßfeller, S. and Strauß, E. (1998): A guide for companies in the agro-food sector to communicate on genetic eengineering to the public, Luxembourg: Office for Official Publications of the EC, European Commission: Document EUR, 18359 EN.

Messner, D. (1995): Die Netzwerkgesellschaft. Wirtschaftliche Entwicklung und internationale Wettbewerbsfähigkeit als Probleme gesellschaftlicher Steuerung, Köln: Weltforum Verlag.

National Institute of Science and Technology Policy (NISTEP 1993): The Fifth Technology Forecast Survey - Future Technology in Japan, Tokyo.

Nedeva, M.; Georghiou, L.; Loveridge, D. and Cameron, H. (1996): The use of co-nomination to identify expert participants for Technology Foresight, in: R&D Management, vol. 26 (1996) no. 2, pp. 155-168.

Neske, F. and Wiener, M. (1985) (eds.): Management-Lexikon, Vol III, Augsburg: DBV.

Norusis, M. J. (1993): SPSS® for Windows™ Professional Statistics™ Release 6.0, Chicago.

Nováky, E. (2000): Methodological Renewal in Futures Studies, Contribution to "The Quest for the Futures". A Methodological Seminar in Futures Studies, June 13-15, 2000, Turku, Finnland.

OECD (ed.) (1996): Special issue on Government Technology Foresight Exercises, Science Technology Industry (STI) Review no. 17, OECD, Paris.

Parke, H. W. and Wormell, D. E. W. (1956): The Delphic Oracle, Oxford: Basil Blackwell.

Pickel, G. (1995): Dimensionen religiöser Überzeugungen bei jungen Erwachsenen in den neuen und alten Bundesländern der Bundesrepublik Deutschland. in: Kölner Zeitschrift für Soziologie und Sozialpsychologie, vol. 47, pp. 517-534.

Polanyi, M. (1985): Implizites Wissen (The Tacit Knowledge), Frankfurt: Suhrkamp.

Reger, G. and Schmoch, U. (eds.) (1996): Organisation of Science and Technology at the Watershed. The Academic and Industrial Perspective, Heidelberg: Physica (Technology, Innovation, and Policy 3).

Rowe, G., Wright, G. and Bolger, F. (1991): Delphi - A Reevaluation of Research and Theory, in: Technological Forecasting and Social Change, vol. 30, pp. 235-251.

Sackman, H. (1975): Delphi Critique. Expert Opinion, Forecasting, and Group Process, Toronto, London: Rand Corporation.

Science and Technology Agency (STA, ed.) (1997): White Paper on Science and Technology, Tokyo.

Sherwin, C. W. and Isenson, R. S. (1967): First Interim Report on Project HINDSIGHT (Summary), Washington, DC., Office of the Director of Defense Research an Engineering, 1966; see also Isenson, R. S.: Technological forecasting lessons from project HINDSIGHT, Harvard University's Technology and Management Conference.

Socrates: Phaidros, ca 400 B. C.

Steinmüller, K. (1995): Beiträge zu Grundfragen der Zukunftsforschung, WerkstattBericht des Sekretariats für Zukunftsforschung 2/95, Gelsenkirchen.

Swinbanks, D. (1994): Earthquake 'forecasters' face their critics in Japan, in: Nature, vol. 370, p. 9.

Swinbanks, D. (1995): Kobe disaster divides earthquake researchers, in: Nature, vol. 373, p. 373.

Thorndike, R. M. (1978): Correlational procedures for research, New York.

Tsuchiya, S. (2001): Human Resource Development for Science and Technology-Driven Nation, Special Feature, in: Science & Technology in Japan, no. 76, pp. 2-5.

Tucker, L. R. (1971): Relations of factor score estimates and their use. in: Psychometrik, vol. 36, pp. 427-436.

van den Daele, W. (1994): Technikfolgenabschätzung als politisches Experiment. Diskursives Verfahren zur Technikfolgenabschätzung des Anbaus von Kulturpflanzen mit gentechnisch erzeugter Herbizidresistenz, in: Bechmann, G. and Petermann, Th. (eds.): Interdisziplinäre Technikforschung. Genese, Folgen, Diskurs, Frankfurt, New York: Campus, pp. 111-146.

van den Daele, W. (1996): Rationalitätsgewinn durch diskursive Verfahren?, Holzinger, K. and Weidner, H. (eds.): Alternative Konfliktregelungsverfahren bei der Planung und Implementation großtechnischer Anlagen, Berlin: WZB: Mimeo.

van den Daele, W. (1997): Risikodiskussionen am "Runden Tisch". Partizipative Technikfolgenabschätzung zu gentechnisch erzeugten herbizidresistenten Pflanzen in: Martinsen, R. (Eds.): Politik und Biotechnologie. Die Zumutung der Zukunft Baden-Baden: Nomos, pp. 281-301.

van der Heijden, K. (1997): Scenarios. The art of strategic conversation, Chichester, New York, Brisbane: Wiley.

Webster's New Encyclopedic Dictionary (1993), New York: Webster.

Wikman, M.; Jacobsson, L.; von Schoultz, B. (1992): Attitudes toward reproduction in a nonpatient population, in: Am J Obstet Gynecol, vol. 166, no. 1, pp. 121-126.

Woudenberg, F. (1991): An Evaluation of Delphi, in: Technological Forecasting and Social Change, vol. 40, pp. 131 – 150.

WZB-Mitteilungen (1997): Abfall: Mediation in Berlin vorbereitet, in: WZB-Mitteilungen no. 76, pp. 19-22.

Internet:

www.tekniskframsyn.nu